Basic Calculations for Chemical and Biological Analysis

Second Edition

Bassey J.S. Efiok, Ph.D
National Heart Lung Blood Institute
National Institutes of Health
Bethesda, MD 20892, USA

Etim Effiong Eduok, Ph.D
Department of Chemistry
Xavier University of Louisiana
New Orleans, LA 70125, USA

Copyright © 2000
By AOAC INTERNATIONAL
481 North Frederick Avenue
Suite 500
Gaithersburg, Maryland 20877-2417 USA

Printed in the United States of America

ISBN 0-935584-69-2

Dedication

Dedicated to the memory of my grandfather, Stephen Udo Akpan, and my Father, James Stephen Udo, whose exemplary wisdom, courage, and leadership still inspire me to strive for higher goals (BJSE).

and

To the memory of my mother and first teacher, Obonganwan Ima Effiong Eduok, who was called home, while this book was in preparation. To my daughters Oto-Obong Eduok, Uyuho Eduok and Ekemini Eduok, three terrific students of the the 21st century (EEE).

Table of Contents

Preface

At some stage of almost every laboratory experiment in the chemical and biochemical sciences, calculations are needed. The types of calculations vary, but often include basic tasks. How much reagent is required to prepare a given solution? And how much should be added to an assay medium to yield a given concentration? How do you convert the experimental data into concentrations, amounts, and activities? And interconvert the units of measure? And how do you calculate chemical composition and properties, as well as the variables and constants for chemical reactions?

Students, laboratory technicians and research scientists are generally assumed to have learned how to perform these calculations from their general or analytical chemistry courses. From our experience as instructors, however, we observed that they either had forgotten the general principles by which quantitative problems are solved or had developed techniques that were not only complex but also failed to give the correct answer.

Our approach can be described in a single word: Simplicity! This book is advanced and practical enough for even experienced scientists in the chemical and biomedical sciences, yet simple enough for a high school senior. Indeed, anyone with a limited education in the chemical sciences will find this book useful. It can be used by college chemistry instructors as a supplemental text to train students how to master basic quantitative problems. Or, by industry laboratory supervisors who can use it as a direct source of materials to train their technical staff on proficiency testing.

The first edition of this book, released in 1993, was a success, with over several thousand copies sold. In this second edition, Professor Etim Eduok and I have joined forces to revise the first edition and develop three new chapters. Our objective was to broaden the scope of the book without compromising its simplicity, and to bring the original content in line with current trends in teaching.

The first eight chapters have wide application in the chemical and biomedical laboratories. Calculations are included for reagent concentration, amount and preparation; chemical reactions and stoichiometry; properties of gases and colligative properties of solutions; acid-base equilibria and pH and buffer preparation; spectrophotometry and its quantitative applications, enzyme assays, and activity; and the quantitative aspects of radioactivity. Chapter eight focuses on the treatment of experimental data using spreadsheets and Laboratory Information Management Systems.

Each chapter includes (1) concise descriptions and definitions for the basic principles (fundamental to understanding the subsequent equations); (2) derivation of basic equations or concepts used for calculations; (3) relevant techniques and their applications; and (4) practical examples that illustrate how the basic equations are used to solve a wide range of common quantitative problems. Most of the examples are real problems. Even the devised examples have been carefully developed to closely mimic real problems.

To ensure that the main text is concise and easy to understand, the derivation of some fundamental equations have been placed in an appendix. Relevant materials, such as instrumentation, principles of measurement, physical/chemical constants, SI units tables, and rec-pes for preparing common laboratory buffers, are also included in the appendices. Finally, quick reference guides to abbreviations, SI units, and equations and an index to the practical examples are provided.

We are very grateful to Todd Jones of the Texas Agricultural Experiment Station, Texas A&M University, College Station, for writing the chapter on spreadsheets and Laboratory Information Management Systems. We are also very grateful to the following: Professor Dale A. Webster of the Department of Biology, Illinois Institute of Technology, Chicago, for the encouragement and inspiration that led to the first edition; Patricia A. Cunniff, Professor of Chemistry, Prince George's Community College, Largo, Maryland, and Peter Spare of the Department of Chemistry Labs, McKellar General Hospital, Ontario, Canada, for their efforts in reviewing the first edition; and Krystyna A. McIver, Director of Publications and other staff of AOAC INTERNATIONAL for their assistance.

Most importantly, we express our sincerest gratitude to you, the thousands of students, technicians, scientists, instructors and supervisors worldwide, who have used this book in the past. We hope that you will try out the second edition, which has been written to bring you more benefits. We invite all others to begin to use this book and experience its many benefits.

Finally, we invite and welcome your comments and suggestions.

Bassey J.S. Efiok, Ph.D
Molecular Hematology Branch
NHLBI, National Institutes of Health
Building 10, Room 7D-18
Bethesda, MD 20892, USA

Etim E. Eduok, Ph.D
Department of Chemistry
Xavier University of Louisiana
7325 Palmetto Street
New Orleans, LA 70125-1098, USA

Quick Reference to Abbreviations and Symbols

A	absorbance
at. wt	atomic weight
a	ionic activity
bp	base pairs
c	concentration
C	centigrade temperature unit
Ci	curie
concn	C concentration
cpm	count per minute
D	Dalton
dpm	disintegrations per minute
E	extinction coefficient
e	base of natural logarithm
equiv	equivalent
equiv wt	equivalent weight
fw	formula weight
g	gram
HOAc	acetic acid
I	ionic strength
IU	international unit
k	rate constant
K	kelvin temperature unit
K_b	molal boiling point elevation constant
K_f	molal freezing point depression constant
L	liter
LIMS	Laboratory Information Management Systems
ℓ	length of light path
log	logarithm to the base 10
M	molarity
m	molality
min	minute
mol	mole

Mol wt	molecular weight
N	Normal (concentration)
N	Avogadro's number
n	number of moles
n	number of equivalents per mole
OAc^-	acetate ion
P	pressure
P	product (of enzyme reaction)
R	universal gas constant ($0.08206 \, L \cdot atm \cdot K^{-1} \cdot mol^{-1}$)
S	second
S	substrate (for enzyme)
sp act.	specific activity
sp vol	specific volume
sp gr	specific gravity
STP	standard temperature and pressure
t	time
V (vol)	volume
wt	weight
w/w	weight-to-weight
w/v	weight-to-volume
Z	number of charge on ion
ΔT_b	boiling point elevation
ΔT_f	freezing point depression
γ	activity coefficient
ρ	density
$\in$	molar extinction
μ	micro
η	number of empirical formula units
χ	mole fraction
Π	osmotic pressure
$\approx$	approximately equal to
[ ]	molar concentration of substance in bracket

Quick Reference to SI Units*

Gram (g)		
Unit	Symbol	Equivalent
picogram	pg	$1\ pg = 10^{-12}$
nanogram	ng	$1\ ng = 10^{-9}$
microgram	μg	$1\ \mu g = 10^{-6}$
milligram	mg	$1\ mg = 10^{-3}$
kilogram	kg	$1\ kg = 10^{3}$
Mole (mol)		
picomole	pmol	$1\ pmol = 10^{-12}$
nanomole	nmol	$1\ nmol = 10^{-9}$
micromole	μmol	$1\ \mu mol = 10^{-6}$
millimole	mmol	$1\ mmol = 10^{-3}$
Liter (l or L)		
nanoliter	nL	$1\ nL = 10^{-9}$
microliter	μL	$1\ \mu L = 10^{-6}$
milliliter	mL	$1\ mL = 10^{-3}$
Curie (Ci)		
curie	Ci	$1\ Ci = 2.22 \times 10^{-12}\ dpm$
millicurie	mCi	$1\ mCi = 10^{-3}\ Ci$
microcurie	μCi	$1\ \mu Ci = 10^{-6}\ Ci$

Note: cpm = dpm × % efficiency

*See Appendix B and Table 7.1 for detailed lists.

Quick Reference to Selected Equations

Equations for Calculating Amounts and Concentrations of Reagents (see Quick Reference to Abbreviations and Symbols on page xiii)

$$\text{mol} = \frac{\text{wt (g)}}{\text{mol wt}}$$

$$\text{equiv} = \frac{\text{wt (g)}}{\text{equiv wt}} \text{ or equiv} = \text{mol} \times n$$

$$\text{M} = \frac{\text{mol}}{\text{L}} \text{ or M} = \frac{\text{wt}}{\text{mol wt} \times \text{L}}$$

$$\text{N} = \frac{\text{equiv}}{\text{L}} \text{ or N} = \frac{\text{wt}}{\text{equiv wt} \times \text{L}}$$

Also,

$$\text{N} = n\text{M}$$

$$\rho = \frac{\text{wt}}{\text{vol}}$$

$$\text{sp gr} = \frac{\rho_{\text{sample}}}{\rho\text{H}_2\text{O}} = \rho_{\text{sample}} \text{ at } 4°\text{C}$$

$$\% \text{ (w/w)} = \frac{\text{g of analyte in sample}}{\text{g of sample}} \times 100\%$$

$$\% \text{ (w/v)} = \frac{\text{g of analyte in sample}}{\text{mL of sample}} \times 100\%$$

$$\text{ppm (w/w)} = \frac{\text{g of analyte in sample}}{\text{g of sample}} \times 10^6 \text{ ppm}$$

$$\text{ppm (w/v)} = \frac{\text{g of analyte in sample}}{\text{mL of sample}} \times 10^6 \text{ ppm}$$

If a volume (V_1) of a reagent (concn C_1) is added to a diluent to yield a final volume (V_2) and a final concentration (C_2), then,

$$\text{Dilution factor} = \frac{C_1}{C_2} = \frac{V_2}{V_1}$$

$$C_1V_1 = C_2V_2 \quad \text{or} \quad M_1V_1 = M_2V_2$$

$$V_1 = \frac{C_2 V_2}{C_1} \quad \text{or} \quad V_1 = \frac{M_2 V_2}{M_1}$$

Equations for Calculating the Weight or Volume of a Reagent Needed to Prepare a Given Solution

For molar concentration,

$$\text{wt(g)} = \text{mol wt (g)} \times M \times L$$

When correcting for % purity,

$$\text{wt(g)} = \text{mol wt (g)} \times M \times L \times \frac{100}{\text{purity of reagent (\%)}}$$

For % and ppm concentration,

$$\text{wt (g)} = \frac{\% \, (\text{w/v}) \times \text{mL of solution}}{100\%}$$

$$\text{wt (g)} = \frac{\text{ppm (w/v)} \times \text{mL of solution}}{10^6 \, \text{ppm}}$$

For liquid reagents,

$$V_1 = \frac{C_2 V_2}{C_1}$$

where V_1 is the volume of the stock reagent which, when diluted to a final volume V_2, yields a desired concentration C_2; C_1 is the concentration of the stock reagent. The units of C_1 and C_2 must be identical, and those of V_1 and V_2 must also be identical.

Equations for Enzyme Activity, Units, and Kinetic Constants

$$\text{activity} = \frac{\text{amount of substrate converted}}{t}$$

or

$$\text{activity} = \frac{\text{amount of product liberated}}{t}$$

$$\text{sp act.} = \frac{\text{amount of substrate converted}}{t \times \text{mg of protein in assay}}$$

or

$$\text{sp act.} = \frac{\text{amount of product liberated}}{t \times \text{mg of protein in assay}}$$

1 IU $= 1$ μmol/min

$$\text{Turnover number} = \frac{\text{mol of substrate catalyzed}}{\text{mol of enzyme} \times \text{s}}$$

In a Lineweaver-Burke plot,

$$V_{max} = \frac{1}{\text{y intercept}}$$

$$K_m = -\frac{1}{\text{x intercept}}$$

$$\frac{K_m}{V_{max}} = \text{slope}$$

Equations Relating to Chemical Equations and Stoichiometry

$$\eta = \frac{\text{molar mass of molecule}}{\text{mass of empirical formula}}$$

$$\% \text{ yield} = \frac{\text{actual yield}}{\text{theoretical yield}} \times 100\%$$

Equations Dealing With the Properties of Gases and the Colligative Properties of Solutions

$$\text{Boyle's Law;} \quad PV = \text{constant}$$

$$\therefore P_1 V_1 = P_2 V_2$$

$$\text{Charles' Law;} \quad \frac{V}{T} = \text{constant}$$

$$\therefore \frac{V_1}{T_1} = \frac{V_2}{T_2}$$

$$\text{Gay-Lusac's Law;} \frac{P}{T} = \text{constant}$$

$$\therefore \frac{P_1}{T_1} = \frac{P_2}{T_2}$$

$$\text{Combined Gas Law;} \quad \frac{PV}{T} = \text{constant}$$

$$\therefore \frac{P_1 V_1}{T_1} = \frac{P_2 V_2}{T_2}$$

$$\text{Ideal Gas Law; } PV = nRT$$

$$\text{Dalton's Law of Partial Pressures; } P_{TOTAL} = \frac{n_1 RT}{V} + \frac{n_2 RT}{V} + \frac{n_3 RT}{V} + \ldots\ldots$$

$$\chi_A = \frac{n_A}{n_A + n_B}$$

$$P_A = \chi_A P_{TOTAL}$$

$$\Delta T_b = K_b m_{solute}; \ \Delta T_f = K_f m_{solute}$$

$$\Pi = MRT$$

$$P_{solvent} = x_{solvent} P^{\circ}_{solvant}$$

Other Commonly Used Equations

$$\text{Beer-Lambert Equation:}$$
$$A = E\ell C$$

Radioactivity remaining [N(t)] after time (t) has elapsed:
$$N_t = N_o e^{-(0.693t)/t_{1/2}}$$

Ionic strength:
$$I = \tfrac{1}{2} \sum M_i Z_i^2$$

Chapter 1

Reagent Quantitation: Calculating Amounts and Concentrations; Preparation Guidelines

This chapter reviews the basic chemical concepts and units used to quantitate reagents (*1, 2*). Equations for calculating the amounts and concentrations of reagents are derived immediately after the relevant chemical concept has been explained. A guideline on reagent preparation is then given, and, finally, practical examples addressing various aspects of calculating for reagent preparation, concentrations, and amounts are presented.

1.1. Review of Basic Atomic and Molecular Concepts

A. Atomic Mass Unit (amu)

By international agreement, 1 amu is defined as one-twelfth the mass of an atom of carbon-12 (^{12}C). When converted to grams, 1 amu = 1.66×10^{-24} g.

B. Mole (mol)

Weighing out or counting individual atoms or molecules is impossible because they are extremely small. The unit of measure, the mole, was introduced to enable scientists to weigh out the amount of any chemical substance that contains a given number of atoms, molecules, or other chemical particles. One mole of atoms is the number of atoms in exactly 12 g of ^{12}C, which is 6.022×10^{23}. This number 6.022×10^{23} is called Avogadro's number. In a broader sense, 1 mole of any substance contains 6.022×10^{23} units. One mole of a chemical element or compound contains 6.022×10^{23} atoms or molecules, respectively, and its weight in grams is equal to the numerical value of the **atomic weight (at. wt)** or **molecular weight (mol wt)** (see sections D and F). For example, the atomic weight of calcium is 40.08 amu. One mole of calcium weighs 40.08 g and contains 6.022×10^{23} atoms. It follows that **the number of moles in a given weight of a substance can be calculated using equation 1, and the weight of a given amount of moles can be calculated using equation 2, below:**

By definition,

$$\text{mol} = \frac{\text{wt (g)}}{\text{mol wt (g / mol)}} \tag{1}$$

$$\therefore \text{wt (g)} = \text{mol} \times \text{mol wt (g / mol)} \tag{2}$$

If the substance is an element, molecular weight is substituted with atomic weight.

Examples: See problems 1, 2a, 5, and 10b in section 1.6.

C. Avogadro's Number (N)

The constant, 6.022×10^{23} atoms/mole or 6.022×10^{23} molecules/mole (see above), is defined as Avogadro's number (N). We can generally say that one mole of something consists of 6.022×10^{23} units of that substance, just as a dozen of eggs is 12 eggs. Therefore, for an element,

$$N = 6.022 \times 10^{23} \text{ atoms} = 1 \text{ mol} = 1 \text{ g-at. wt}$$

For a compound,

$$N = 6.022 \times 10^{23} \text{ molecules} = 1 \text{ mol} = 1 \text{ g-mol wt}$$

Examples: See problems 3 and 4 in section 1.6.

D. Atomic Weight (at. wt) and Gram-Atomic Weight (g-at. wt)

Atomic weight is the weight (in amu) of one atom of an element. It comprises the weight of the protons and neutrons in the atomic nucleus. The gram-atomic weight of any element is the weight (in grams) of 6.022×10^{23} atoms or 1 mole of that element. The gram-atomic weight is numerically equal to the atomic weight. For example, the atomic weight of sulfur (S) is 32.06 amu; the g-at. wt is 32.06 g. The 32.06 g is calculated as follows: There are 6.022×10^{23} atoms/mole S. Therefore, the mass (amu) of 1 mole of S is

$$1 \text{ mol S} \times \frac{6.022 \times 10^{23} \text{ atoms S}}{1 \text{ mol S}} \times \frac{32.06 \text{ amu}}{1 \text{ atom}} = 1.9307 \times 10^{25} \text{ amu}$$

This is then converted from amu to grams:

$$1.9307 \times 10^{25} \text{ amu} \times \frac{1.6606 \times 10^{-24} \text{ g}}{1 \text{ amu}} = 32.06 \text{ g}$$

E. Molecular Formula

The atoms that make up one molecule of a compound constitute the chemical or molecular formula of that compound. For example, the formula of a molecule of glucose, which consists of 6 atoms of carbon, 12 atoms of hydrogen, and 6 atoms of oxygen, is $C_6H_{12}O_6$. The atoms in this formula are covalently linked and do not dissociate in solution. Therefore, $C_6H_{12}O_6$ exists as a discrete molecule. In contrast, discrete molecules of some compounds exist only in concept because the molecules are formed by associated ions. For example, the formula of magnesium sulfate is $MgSO_4$. In solution, it dissociates into one magnesium ion (Mg^{2+}) and one sulfate ion (SO_4^{2-}). A discrete molecule of magnesium sulfate or any other dissociable compound, therefore, exists only in concept. For details on these concepts, see chapter 2.

F. Molecular Weight (mol wt) and Gram-Molecular Weight (g-mol wt)

The molecular weight of a compound is the sum of the atomic weight (in amu) of all the atoms that make up one molecule of the compound. The molecular weight (in g) of any compound contains 6.022×10^{23} molecules or 1 mole of the compound. For this reason, the molecular weight, expressed in grams, is termed gram-molecular weight. The gram-molecular weight of a compound is numerically equal to its molecular weight, and this can be calculated similar to the example given in section D, above, for the g-at. wt of sulfur. Another commonly used name is **molar mass** (see chapter 2).

G. Formula Weight (fw)

This unit is used in place of molecular weight to designate the weight of the formula of a compound that does not exist as a discrete molecule (section E). It is defined as the sum of the atomic weights of all the elements that comprise the chemical formula of the compound. The formula weight (g) of any compound contains 6.022×10^{23} molecules or 1 mole of the compound.

H. Dalton (D)

This unit is used to report the molecular or atomic mass of substances; 1 Dalton is equal to 1 amu, or one-twelfth the mass of one atom of ^{12}C (see section 1.2A). Although the number of Daltons for a molecule is equivalent to the molecular weight, the Dalton is generally reserved for reporting the masses of macromolecular substances such as proteins.

I. Equivalent Weight (equiv wt)

This is the weight of an acid or a base containing 1 mole of replaceable H^+ or OH^-, respectively, or the weight of a redox compound that contains 1 mole of exchangeable electrons, or the weight of an ionic substance carrying 1 mole of ions. Gram-equivalent weight is used when equivalent weight is expressed in grams. **The equivalent weight of a substance is calculated using equation 3, below:**

$$\text{equiv wt} = \frac{\text{mol wt}}{n} \qquad (3)$$

where n is the number of **equivalents** per mole (or the number of replaceable H^+ or OH^-, exchangeable electrons, or charge, per molecule or ion of the substance). When $n = 1$, then **equiv wt = mol wt.**

Examples: See problems 18 and 19 in section 1.6.

J. Equivalent (equiv)

This is an operational term that defines the gram-equivalent weight of an acid, a base, an electron transferring substance, or an ion. One equivalent of a compound contains 1 g-equiv wt of the compound. **The number of equivalents in a given weight of a substance is calculated using equations 4 and 6, and the weight containing a given number of equivalents is calculated using equation 5, below.** By definition,

$$\text{equiv} = \frac{\text{wt (g)}}{\text{equiv wt (g)}} \qquad (4)$$

$$\therefore \text{wt (g)} = \text{equiv} \times \text{equiv wt} \qquad (5)$$

Another useful equation can be derived by combining equations 3 and 5:

$$\text{mol wt} = \frac{\text{wt}}{\text{mol}} = \text{equiv wt} \times n$$

$$\text{or} \frac{\text{wt}}{\text{equiv wt}} = n \times \text{mol}$$

Since, by equation 4,

$$\text{equiv} = \frac{\text{wt}}{\text{equiv wt}}$$

it follows that:

$$\text{equiv} = n \times \text{mol} \tag{6}$$

Again, n is the number of equiv/mol.

Example: See problem 19 in section 1.6.

1.2. Calculating Concentrations Based on Mole and Equivalent

A. Molarity (M)

This concentration unit is defined as the number of moles of a substance per liter (**L**) of solution. For example, a 2 M solution of NaCl contains 2 moles of NaCl/L. **The molarity of a substance in solution is calculated using equations 7 and 8, and the weight of a substance needed to prepare a solution of a given volume (in liters) and molarity is calculated using equation 9, below.** By definition,

$$M = \frac{\text{mol}}{\text{L}} \tag{7}$$

By substituting equation 1 into equation 7,

$$M = \frac{\text{wt (g)}}{\text{mol wt} \times \text{L}} \tag{8}$$

$$\therefore \text{wt (g)} = M \times \text{mol wt} \times \text{L} \tag{9}$$

Examples: See problems 2b, 6, 7, 8, 10c, 11, 12, 13c, and 14 in section 1.6.

When the purity of a reagent is less than 100%, the calculated weight is corrected by multiplying with the factor:

$$\frac{100}{\text{purity of reagent (\%)}}$$

Equation 9 then becomes:

$$wt\ (g) = (mol\ wt \times M \times L) \times \frac{100}{purity\ of\ reagent\ (\%)} \qquad (10)$$

Examples: See problems 12 and 14 in section 1.6.

B. Molality (m)

This is a concentration unit defined as the number of moles of a solute per kilogram of solvent; i.e., a 1 molal solution of glucose is prepared by adding 180.16 g (1 mol) of glucose to 1 kg of H_2O. **The molality of a substance in a solution is calculated using equations 11 and 12, and the weight of a substance needed to prepare a solution with a given molality and solvent weight (in kg) is calculated using equation 13, below.** By definition,

$$molality = \frac{mol}{kg\ of\ solvent} \qquad (11)$$

By substituting equation 1 into equation 11,

$$molality = \frac{wt\ (g)}{mol\ wt \times kg\ of\ solvent} \qquad (12)$$

$$\therefore wt\ (g) = molality \times mol\ wt \times kg\ of\ solvent \qquad (13)$$

Example: See problem 9 in section 1.6.

C. Formality (f)

This concentration unit is defined as the number of gram formula weight of a substance per liter of solution. The value is the same as molarity.

D. Osmolarity (osM)

This concentration unit is defined as the moles of particles per liter of solution. For substances that do not dissociate in solution, the osmolar concentration is equal to the molar concentration. For ionizable substances, the osmolar concentration is equal to z times the molarity, where z is the number of ions produced per molecule upon ionization. Thus,

$$osM = z \times M \qquad (14)$$

$$and\ milliosmolarity\ (mosM) = z \times mM \qquad (15)$$

where mM is 1000th of 1 M, and mosM is 1000th of 1 osM. Because osmolarity is too large a unit, milliosmolarity is more widely used.

E. Normality (N)

This concentration unit is defined as the number of equivalents of a substance per liter of solution. For example, a 0.5 N solution of acetic acid contains 0.5 equiv/L. **Normality is calculated using equations 16 and 17, and the weight of a substance needed to prepare a solution of a given volume (in liters) and normality is calculated using equation 18, below:**

$$N = \frac{\text{equiv}}{L} \qquad (16)$$

By substituting equation 4 into equation 16,

$$N = \frac{\text{wt (g)}}{\text{equiv wt (g)} \times L} \qquad (17)$$

$$\text{wt (g)} = N \times \text{equiv wt (g)} \times L \qquad (18)$$

Examples: See problems 20–22 in section 1.6.

F. Molarity Related to Normality

From equations 6 and 7,

$$\text{equiv} = n \times \text{mol}$$

$$\text{mol} = M \times L$$

where equiv and mol represent the number of equivalents and the number of moles, respectively. Combining the two equations yields:

$$\text{equiv} = n \times M \times L \qquad (19)$$

From equation 16,

$$\text{equiv} = N \times L \qquad (20)$$

Combining equations 19 and 20 yields

$$N \times L = n \times M \times L$$

$$\therefore N = nM \qquad (21)$$

where n is the number of equivalents per mole.

Examples: See problems 20 and 21 in section 1.6.

1.3. Calculating Concentrations Based on Weight

Weight-based concentration units in frequent use are percent (%), parts per million (ppm), parts per billion (ppb), density (ρ), specific gravity (sp gr), and specific volume (sp vol). Percent may be defined as parts per 100 units, and the equations for calculating the various types of percent concentration are derived from the following general expression:

$$\% \text{ analyte} = \frac{\text{wt of analyte in sample}}{\text{wt or vol of sample}} \times 100\% \qquad (22)$$

where analyte refers to the substance being determined, and sample refers to the material that contains the analyte and is being analyzed. If the **amount of sample** (denominator) is in a weight unit, the calculated **percent analyte** is designated % (w/w); if it is in a volume unit, the percent analyte is designated % (w/v). To derive similar expressions for calculating parts per million and parts per billion, the 100% is replaced by 10^6 ppm or 10^9 ppb, respectively.

Specific equations for calculating each of the weight-based concentrations, and a corresponding weight of reagent needed to prepare a solution of a given concentration are derived below.[1]

A. Percent (Weight/Weight), or % (w/w)

This is the grams of pure analyte in 100.00 g of sample. For example, a 37% (w/w) solution of commercial concentrated HCl contains 37.00 g of pure HCl in 100.00 g of solution. **The % (w/w) of an analyte in a sample is calculated using equation 23 below:**

$$\% \text{ (w/w)} = \frac{\text{g of analyte in sample}}{\text{total g of sample}} \times 100\% \qquad (23)$$

For the units in these equations to cancel out, the %, ppm, or ppb units in the equations must be substituted with their gram and/or volume equivalents. For example, when dealing with % (w/w), replace (% analyte) with (g of analyte/100.00 g of sample), and the % to the right of the equation with (g of sample/100.00 g of sample); and when dealing with % (w/v), replace (% analyte) with (g of analyte/100.0 mL of sample), and the % to the right with (mL of sample/100.0 mL of sample). See problems 14 and 15 in section 1.6.

B. Percent (Weight/Volume), or % (w/v)

This is the grams of pure analyte in 100.0 mL of solution. For example, a 10% (w/v) sucrose solution contains 10.00 g of sucrose per 100.0 mL. **The % (w/v) is calculated using equation 24, and the grams of pure analyte needed to prepare a solution with a given volume (mL) and percent concentration are calculated using equation 25, below:**

$$\% \text{ (w/v)} = \frac{\text{g of analyte in sample}}{\text{mL of sample}} \times 100\% \qquad (24)$$

$$\therefore \text{ g of analyte} = \frac{\% \text{ (w/v)} \times \text{ mL of sample}}{100\%} \qquad (25)$$

Examples for Equations 23–25: See problems 13 and 15 in section 1.6.

C. Milligram Percent (mg %)

This is the milligrams of pure analyte in 100.0 mL of solution. For example, a 20 mg % solution of NaCl contains 20 mg of NaCl in 100.0 mL of the solution. **The mg % is calculated using equation 26, and the milligrams of pure analyte needed to prepare a solution with a given volume (mL) and mg % concentration is calculated using equation 27, below:**

$$\text{mg \%} = \frac{\text{mg of analyte in sample}}{\text{mL of sample}} \times 100\% \qquad (26)$$

$$\therefore \text{mg of analyte} = \frac{\text{mg \%} \times \text{mL of sample}}{100\%} \qquad (27)$$

D. Parts per Million (Weight/Weight), or ppm (w/w)

This is the grams of pure analyte in 10^6 g of sample. For example, if trace Na^+ in solid $MgCl_2$ is stated as 10 ppm, then 10.00 g of Na^+ is contained in every 10^6 g of $MgCl_2$. **The ppm (w/w) is calculated using equation 28, below:**

$$\therefore \text{ppm (w/w)} = \frac{\text{g of analyte in sample}}{\text{g of sample}} \times 10^6 \text{ ppm} \quad (28)$$

Examples: See problems 16 and 17 in section 1.6.

E. Parts per Million (Weight/Volume), or ppm (w/v)

This is the grams of pure analyte in 10^6 mL of sample. If, for example, the concentration of Na^+ in a standard solution is stated as 1000 ppm, the solution contains 1000.00 g of $Na^+/10^6$ mL. **The ppm (w/v) is calculated using equation 29, and the grams of pure analyte needed to prepare a solution with a given volume (mL) and ppm (w/v) concentration is calculated using equation 30, below:**

$$\therefore \text{ppm (w/v)} = \frac{\text{g of analyte in sample}}{\text{mL of sample}} \times 10^6 \text{ ppm} \quad (29)$$

$$\therefore \text{g of analyte} = \frac{\text{ppm (w/v)} \times \text{mL of sample}}{10^6 \text{ ppm}} \qquad (30)$$

Examples: See problems 16 and 17 in section 1.6.

F. Parts per Billion (Weight/Weight), or ppb (w/w)

This is the grams of pure analyte in 10^9 g of sample. Thus, if the concentration of trace Pb^{2+} in solid $CaCl_2$ is stated as 50 ppb, then 10^9 g of the $CaCl_2$ contains 50.00 g of Pb^{2+}. **The ppb (w/w) is calculated using equation 31, below:**

$$\text{ppb (w/w)} = \frac{\text{g of analyte in sample}}{\text{g of sample}} \times 10^9 \text{ ppb} \qquad (31)$$

G. Parts per Billion (Weight/Volume), or ppb (w/v)

This is the grams of pure analyte in 10^9 mL of sample. Thus, if the concentration of trace K^+ in concentrated H_2SO_4 is stated as 100 ppb, the solution contains 100.00 g of $K^+/10^9$ mL. **The ppb (w/v) is calculated using equation 32, and the grams of pure analyte needed to prepare a solution with a given mL and ppb (w/v) concentration is calculated, below, using equations 32 and 33, respectively:**

$$\text{ppb (w/v)} = \frac{\text{g of analyte in sample}}{\text{mL of sample}} \times 10^9 \text{ ppb} \qquad (32)$$

$$\therefore \text{ g of analyte } = \frac{\text{ppb } (\text{w/v}) \times \text{mL of sample}}{10^9 \text{ ppb}} \qquad (33)$$

H.   Density (ρ) and Specific Gravity (sp gr)

Density is the quantity of mass per unit volume of a substance. It is calculated using equation 34, with weight used to approximate mass.

$$\rho = \frac{\text{wt}}{\text{vol}} \qquad (34)$$

Specific gravity is the density of a fluid relative to that of H_2O:

$$\text{sp gr } = \frac{\rho_{\text{sample}}}{\rho H_2 O} \qquad (35)$$

Substituting 1 g/mL for the density of H_2O at 4°C into equation 35,

$$\text{sp gr } = \frac{\rho_{\text{sample}} \ (\text{g / mL})}{1 \ (\text{g / mL})} \qquad (36)$$

Therefore, the specific gravity of a fluid is numerically equal to its density at a temperature of 4°C.

Examples: See problems 12 and 20 in section 1.6.

I.   Specific Volume (sp vol)

This is the volume occupied by a unit weight of a solute when dissolved. It is calculated using equation 37, below:

$$\text{sp vol } = \frac{\text{vol of dissolved solute}}{\text{wt of solute}} = \frac{1}{\rho} = \rho^{-1} \qquad (37)$$

Therefore, the specific volume of a solute is equal to the inverse of its density.

1.4.   Dilution Factor in Concentration and Volume Calculations

Often, one seeks the volume of a stock reagent that should be added to a known volume of diluent or assay medium to yield a desired concentration, or the final concentration of the reagent after adding a known volume to the diluent or assay medium. An equation relating the various parameters is derived as follows: Let V_1 and C_1, respectively, be the needed volume and the needed concentration of the stock reagent or solution; and V_2, C_2, the final volume and desired concentration, respectively, of the diluted reagent. We can also use M_1 and V_1 to denote respectively the needed molarity and volume of the stock reagent or solution while M_2 and V_2 denote the final and desired molarity and volume, respectively, of the diluted reagent or solution. Then,

$$\text{dilution factor} = \frac{C_1}{C_2} = \frac{V_2}{V_1} \tag{38}$$

$$\therefore C_1V_1 = C_2V_2 \tag{39}$$

$$\text{or } M_1V_1 = M_2V_2 \tag{40}$$

Any of the parameters in equation 39 and 40 may be calculated, provided that the other three are known. In the calculations, C_1 and C_2 or M_1 and M_2 must have identical units; likewise, V_1 and V_2 must have identical units.

Examples: See problems 10d, 12, 20c, 22, and 23c in section 1.6.

1.5. Preparing Reagents

The steps involved in reagent preparation are outlined below and include references to relevant equations and examples in the text.

(1) Examine the purity, formula weight, and mole ratio data for the stock reagent

If the percent impurity of the stock reagent is high, apply a correction factor in step 2, below. This ensures that the concentration of the pure solute in the final solution is not significantly lower than the expected value (see problems 12 and 14 in section 1.6. Use also the Five-step problem solving format, below). If you have to calculate the formula weight, make sure that it contains the atomic weight of all the elements in the chemical formula (including bound H_2O and ions; see problem 11 for example). When preparing a solution of a substance that is part of a compound, and 1 mole of the compound contains more than 1 mole of the substance, determine what molarity of the compound will yield the desired molarity of the substance. Use this value in step 2, below. For example, when preparing a 100 mM solution of OAc^- using $Mg(OAc)_2$, base the calculations on 50 mM $Mg(OAc)_2$, because 1 mole of $Mg(OAc)_2$ contains 2 mole of OAc^-. See also problem 8.

(2) Calculate the amount of the stock reagent needed

   (a) Solid reagents

If the stock reagent is a solid, calculate the amount needed according to the information provided in Table 1.1 and then use the five-step format shown below:
   Step 1: Weigh out the g of solute needed.
   Step 2: Add approximately one-half of the required total volume of solvent into a volumetric flask or graduated cylinder.[2]
   Step 3: Add the solute in Step 1 to the solvent in Step 2 and mix until it dissolves.
   Step 4: Adjust the volume to the required total volume by adding more solvent.
   Step 5: Mix the solution thoroughly.

Examples: See problems 24 and 25.

   (b) Liquid reagents

If the stock reagent is a liquid, calculate the volume needed, using equation 39 or 40. Use the Five- step format below. **Examples** are given in problems 10d, 12, 20c, 22, 23c, 27, 28, and 29 in section 1.6.

Table 1.1. Equations Used to Calculate Amount of Solid Reagent Needed to Prepare a Given Solution

Unit for concn.	Equation no.	Example problems
mol/L (M)	9	7, 8, 11
mol/L (corrected for % impurity)	10	12, 14
mol/kg (Molarity)	11	9
equiv/L (N)	12	18, 20c, 22
% (w/v)	15	15
mg %	16	—
ppm (w/v)	17	—
ppb (w/v)	18	—
M or N (acid solutions)[a]	10, 12, 35	20–22
g/mL	g needed = C (g/mL) × V (mL)[b]	

[a] See step 5.
[b] C and V denote concentration and volume, respectively.

Step 1: Calculate the needed mL of the stock solution using equation 39 or 40. Measure out the calculated volume (V_1).

Step 2: Subtract V_1 from the needed total volume, V_f. Divide the resulting volume by 2. $[V_f - V_1]/2$=volume of solvent to be added to flask.[3] Add this volume of solvent to a volumetric flask or graduated cylinder.

Step 3: Add the measured amount of stock solution in step 1 to the solvent in step 2.

Step 4: Adjust the volume to the required total volume by adding more solvent.

Step 5: Mix the solution thoroughly.

Caution: To protect against accidental chemical injury, dissolve concentrated acid and base solutions under a hood. Add the acid slowly to the H_2O and stir. Pouring H_2O rapidly into a concentrated acid solution causes a violent reaction that may lead to an explosion! Wear goggles to protect your eyes.

Examples: See problems 26 and 27.

1.6. Practical Examples

PROBLEM 1

(a) If there are 8.0 mg of Na^+ contaminant per liter of H_2O sample, calculate the number of moles of Na^+ per liter. The at. wt of Na is 23.0. (b) Calculate the weight of Na^+ in another sample of H_2O that contains 100 μmol Na^+/L.

SOLUTION:

(a) From equation 1,

$$mol = \frac{wt\ (g)}{mol\ wt\ (g\ /\ mol)}$$

$$\therefore mol\ Na^+\ /\ L = \frac{(0.0080\ g\ Na^+)\ /\ L}{23.0\ g\ /\ mol} = 3.5 \times 10^{-4}\ mol\ /\ L$$

(b) From equation 1,

$$mol = \frac{wt\ (g)}{mol\ wt\ (g\ /\ mol)}$$

$\therefore$ wt of Na^+ / L of H_2O = mol $\times$ mol wt = 1×10^{-4} mol $\times$ 23.0 g / mol

$$= 2.30 \times 10^{-3}\ g$$

$$100\ \mu mol = (100\ \mu mol) \times (1\ mol) / (1 \times 10^6\ \mu mol)$$
$$= 1 \times 10^{-4}\ mol\ (see\ Appendix\ C,\ Table\ C\text{-}6)$$

PROBLEM 2

If there are 8.0 mg of dissolved O_2 per liter of H_2O at 20°C and 1 atm, calculate (a) the number of moles of molecular O_2 and (b) the molarity of dissolved O_2 per liter of H_2O. The mol wt of O_2 is 32.0.

SOLUTION:

(a) Using equation 1,

$$mol = \frac{wt\ (g)}{mol\ wt\ (g\ /\ mol)}$$

$$\therefore mol\ O_2\ /\ L = \frac{(0.0080\ g\ O_2)\ /\ L}{32.0\ g\ /\ mol} = 2.5 \times 10^{-4}\ mol\ /\ L$$

(b) Using equation 7,

$$M\ of\ dissolved\ O_2 = \frac{mol}{L} = \frac{2.5 \times 10^{-4}\ mol}{1\ L} = 2.5 \times 10^{-4}\ M\ or\ 250\ \mu M$$

PROBLEM 3

Calculate the number of potassium (K) atoms in a 1.00 μg sample of pure KCl. The at. wt of K is 39.1 and the mol wt of KCl is 74.6.

SOLUTION:

$$Ratio\ of\ K\ in\ 1\ mol\ of\ KCl = \frac{at.\ wt\ of\ K}{mol\ wt\ of\ KCl} = \frac{39.1}{74.6}$$

$$\therefore g\ of\ K\ in\ 1.00\ \mu g\ KCl = \frac{39.10\ g\ of\ K}{74.60\ g\ of\ KCl} \times 1.00 \times 10^{-6}\ g\ of\ KCl$$

$$= 5.24 \times 10^{-7}\ g\ of\ K$$

Number of atoms in 1 mol of K (39.1 g) = 6.022×10^{23} atoms

$\therefore$ number of atoms in 5.24×10^{-7} g of K = mol of K $\times$ 6.022×10^{23} atoms / mol

$$= \frac{5.24 \times 10^{-7} \text{ g}}{39.10 \text{ g}} \times \frac{6.022 \times 10^{23} \text{ atoms}}{1 \text{ mol}} = 8.07 \times 10^{15} \text{ atoms}$$

PROBLEM 4

Calculate the number of molecules of ampicillin per µL of a 10.0 nM solution.

SOLUTION:

1.0 L or 1000.0 mL of the solution contains 10.0 nmol of ampicillin. Therefore, 1.0 µL of the solution contains

$$\frac{10.0 \text{ nmol}}{1000.0 \text{ mL}} \times (1 \times 10^{-3} \text{ mL}) = \frac{1.0 \times 10^{-8} \text{ mol}}{10^{3} \text{ mL}} \times (1 \times 10^{-3} \text{ mL})$$

$$= 1.0 \times 10^{-14} \text{ mol}$$

Number of molecules in 1.0 of ampicillin $= 6.022 \times 10^{23}$ molecules.

$$\therefore \text{ number of molecules in } 1.0 \times 10^{-14} \text{ mol ampicillin}$$
$$= \text{ mol ampicillin} \times 6.022 \times 10^{23} \text{ molecules / mol}$$

$$= \frac{6.022 \times 10^{23} \text{ molecules}}{1.0 \text{ mol}} \times (1.0 \times 10^{-14} \text{ mol / µL}) = 6.0 \times 10^{9} \text{ molecules / µL}$$

PROBLEM 5

For 6.70 g of sodium acetate (NaOAc) and 6.70 g of $Mg(OAc)_2 \cdot 4H_2O$, calculate (a) the moles of each salt, (b) the moles of acetate (OAc^-) in the NaOAc, and (c) the moles of OAc^- in the $Mg(OAc)_2 \cdot 4H_2O$. The formula weights are: NaOAc, 82.03; $Mg(OAc)_2 \cdot 4H_2O$, 214.46.

SOLUTION:

(a) From equation 1,

$$\text{mol} = \frac{\text{wt (g)}}{\text{mol wt (g / mol)}}$$

$$\therefore \text{mol NaOAc} = \frac{6.70 \text{ g}}{82.03 \text{ g / mol}} = 0.082 \text{ mol}$$

$$\text{and mol } Mg(OAc)_2 \cdot 4H_2O = \frac{6.70 \text{ g}}{214.46 \text{ g / mol}} = 0.031 \text{ mol}$$

(b) Moles of OAc^- in each salt can be calculated in two ways:
(i) 1 mol of NaOAc contains 1 mol of OAc^-

$$\therefore \text{mol of } OAc^- \text{ in } 0.082 \text{ mol of NaOAc} = 1 \times 0.082 \text{ mol} = 0.082 \text{ mol}$$

(ii) mol of OAc^- / mol of NaOAc $= \dfrac{59.03}{82.03}$

$$\therefore \text{ g OAc}^- \text{ in 6.70 g NaOAc } = \frac{59.03}{82.03} \times 6.70 \text{ g } = 4.82 \text{ g}$$

Using equation 1,

$$\text{mol OAc}^- \text{ in 6.70 g NaOAc } = \frac{\text{g of OAc}^-}{\text{mol wt of OAc}^-} = \frac{4.82 \text{ g}}{59.03 \text{ g / mol}} = 0.082 \text{ mol}$$

(c) Following the above examples (5b),

(i) 1 mol of $Mg(OAc)_2 \cdot 4H_2O$ contains 2 mol of Oac^-

$$\therefore \text{ mol of OAc}^- \text{ in 0.031 mol of Mg(OAc)}_2 \cdot 4H_2O = 2 \times 0.031 \text{ mol } = 0.062 \text{ mol}$$

(ii) As an exercise, calculate the moles of OAc^- in $Mg(OAc)_2 \cdot 4H_2O$ using method (ii), above (answer = 0.062).

PROBLEM 6

If 6.70 g of sodium acetate (NaOAc) or 6.70 g of $Mg(OAc)_2 \cdot 4H_2O$ is dissolved in a total volume of 0.200 L of H_2O, calculate (a) the molarity of NaOAc, (b) the molarity of $Mg(OAc)_2 \cdot 4H_2O$, and (c) the molarity of OAc^- in each solution. Use the formula weights given in problem 5.

SOLUTION:

The problems can be solved using either equation 7 or 8.

(a) Using equation 8,

$$M = \frac{\text{wt}}{\text{mol wt} \times \text{L}}$$

$$\therefore \text{ M of NaOAc } = \frac{6.70 \text{ g}}{82.03 \text{ g / mol} \times 0.200 \text{ L}} = 0.41 \text{ mol / L}$$

(b) Following the example in (a),

$$\text{M of Mg(OAc)}_2 \cdot 4H_2O = \frac{6.70 \text{ g}}{214.46 \text{ g / mol} \times 0.200 \text{ L}} = 0.16 \text{ mol/L}$$

(c) There is 1.0 mole of OAc^-/mole of NaOAc.

$$\therefore \text{ M of OAc}^- \text{ in NaOAc } = 1.0 \times \text{ M of NaOAc} = 1.0 \times 0.41 \text{ mol / L} = 0.41 \text{ M}$$

There are 2.0 moles of OAc^-/mole of $Mg(OAc)_2 \cdot 4H_2O$.

$$\therefore \text{ M OAc}^- \text{ in Mg(OAc)}_2 \cdot 4H_2O$$

$$2.0 \times \text{ M of Mg(OAc)}_2 \cdot 4H_2O = 2.0 \times 0.16 \text{ mol / L} = 0.32 \text{ M}$$

PROBLEM 7

How many grams of EDTA·$4H_2O$ (fw, 380) are needed to prepare 0.100 L of a 0.020 M solution?

SOLUTION:

Using equation 9,

$$\text{g of EDTA needed} = M \times \text{mol wt} \times L$$
$$= 0.020 \text{ mol} / L \times 380 \text{ g} / \text{mol} \times 0.100 L = 0.76 \text{ g}$$

PROBLEM 8

Prepare 0.100 L of a 0.050 M solution of OAc^- using (a) NaOAc (fw, 82.03), or (b) $Mg(OAc)_2$·$4H_2O$ (fw, 214.46).

SOLUTION:

(a) There is 1.0 mole of OAc^-/mole of NaOAc. Therefore, 0.050 M OAc^- is equivalent to 0.050 M NaOAc. From equation 9,

$$\text{wt} = M \times \text{mol wt} \times L$$

$$\text{g of NaOAc needed} = 0.050 \text{ mol} / L \times 82.03 \text{ g} / \text{mol} \times 0.100 L = 0.41 \text{ g}$$

(b) There are 2.0 mole of OAc^-/mole of $Mg(OAc)_2$·$4H_2O$. Therefore, 0.050 M OAc^- is equivalent to 0.050/2 or 0.025 M $Mg(OAc)_2$·$4H_2O$.

Using equation 9,

$$\text{g of } Mg(OAc)_2 \cdot 4H_2O \text{ needed} = 0.025 \text{ mol} / L \times 214.46 \text{ g} / \text{mol} \times 0.100 L = 0.54 \text{ g}$$

Each salt is then dissolved in about 50.0 mL of H_2O, after which the volume is increased to 100.0 mL with additional H_2O.

PROBLEM 9

Prepare 500.0 mL of a 2.0 molal aqueous solution of KCl.

SOLUTION:

Using equation 13, calculate the weight of KCl needed:

$$\text{wt (g)} = \text{molality} \times \text{mol wt} \times \text{kg of solvent}$$

The mol wt of KCl is 74.55 g/mol, and 500.0 mL of H_2O weighs 0.500 kg.

$$\therefore \text{wt of KCl needed} = 2.0 \text{ mol} / \text{kg} \times 74.55 \text{ g} / \text{mol} \times 0.500 \text{ kg} = 74.5 \text{ g}$$

To prepare the solution, dissolve 74.55 g of KCl in 250.0 mL of H_2O, and then adjust the volume to 500.0 mL with additional H_2O. See section 1.5 for method.

PROBLEM 10

The concentration of ampicillin in a stock solution is 5.00×10^3 µg/mL. Calculate:

(a) µg of ampicillin in 0.100 mL of the solution, (b) µmol of ampicillin in 0.100 mL of the solution, (c) molarity of ampicillin in the solution, and (d) prepare 30.0 mL of a working solution containing 10.0 µg/mL. The fw of ampicillin is 349.41.

SOLUTION:

(a) 1.0 mL of the solution contains 5.00×10^3 µg

$$\therefore 0.100 \text{ mL contains } \frac{5.00 \times 10^3 \text{ µg}}{1.0 \text{ mL}} \times 0.100 \text{ mL}$$

$$= 5.00 \times 10^2 \text{ µg or } 5.00 \times 10^{-4} \text{ g of ampicillin}$$

(b) Using the result in (a), and equation 1,

$$\text{mol of ampicillin} = \frac{g}{g \text{-} \text{mol wt}} = \frac{5.00 \times 10^{-4} \text{ g}}{349.41 \text{ g / mol}} = 1.43 \times 10^{-6} \text{ mol or } 1.43 \text{ µmol}$$

(c) From the result in (b), 0.100 mL (or 1.0010^{-4} L) of the solution contains 1.43×10^{-6} mol. Then, using equation 7,

$$\text{M of ampicillin} = \frac{\text{mol}}{\text{L}} = \frac{1.43 \times 10^{-6} \text{ mol}}{1.00 \times 10^{-4} \text{ L}} = 0.0143 \text{ M or } 14.3 \text{ mM}$$

The molarity can also be calculated using the result from (a) and equation 10:

$$M = \frac{\text{g of ampicillin}}{\text{mol wt} \times \text{L}} = \frac{5.00 \times 10^{-4} \text{ g}}{349.41 \text{ g / mol} \times 1.00 \times 10^{-4} \text{ L}} = 0.0143 \text{ M}$$

(d) Let V_1 be the volume of the stock ampicillin solution needed to prepare the 30.0 mL (V_2) solution, and let C_1 and C_2 be the concentrations of ampicillin in the stock solution and the 30.0 mL solution, respectively. Then, using equation 39,

$$\therefore V_1 = \frac{C_2 V_2}{C_1} = \frac{10.0 \text{ µg / mL} \times 30.0 \text{ mL}}{5.00 \times 10^3 \text{ g / mL}} = 0.0600 \text{ mL or } 60.0 \text{ µL}$$

To prepare the solution, add 60.0 µL of the ampicillin solution to 29.9 mL of diluent.

PROBLEM 11

How many grams of ATP are needed to prepare 0.200 L of a 100.0 µM solution? The anhydrous mole wt of ATP is 487.20. The ATP powder being used has 3.0 moles of H_2O and 2.0 moles of Na^+ per mole.

SOLUTION:

First calculate the formula weight of the ATP:

$$\text{fw} = \text{mol wt of ATP} + 3.0 \text{ mol wt of } H_2O + 2.0 \text{ mol wt of } Na^+$$
$$= 487.20 + (3.0 \times 18.0) + (2.0 \times 23.0) = 587.20$$

Then, calculate the grams of ATP needed, using equation 9:

$$\text{g of ATP} = M \times \text{mol wt} \times L$$
$$= 0.0001 \text{ mol} / L \times 587.20 \text{ g} / \text{mol} \times 0.200 \text{ L} = 0.012 \text{ g}$$

PROBLEM 12

ß-Mercaptoethanol is available as a 98% (w/w) solution (sp gr 1.114). What volume is needed to prepare 0.100 L of a 50.0 mM solution? Assume a fw of 78.13.

SOLUTION:

The volume can be calculated in two ways:

(i) Calculate the grams of ß-mercaptoethanol needed and how many milliliters of the stock solution are equivalent to the calculated weight. Using equation 10,

$$\text{g of } \beta\text{-mercaptoethanol needed} = M \times \text{mol wt} \times L \times \frac{100\%}{\% \text{ purity}}$$

$$= \frac{0.05 \text{ mol} / L \times 78.13 \text{ g} / \text{mol} \times 0.100 \text{ L} \times 100\%}{98\%} = 0.40 \text{ g}$$

Using equation 34,

$$\text{vol} = \frac{\text{wt}}{\rho} = \frac{0.40 \text{ g}}{1.114 \text{ g} / \text{mol}} = 0.4 \text{ mL}$$

(ii) Calculate the molarity of ß-mercaptoethanol in the stock solution. Use it to calculate how many milliliters should be diluted to 100.0 mL to yield 50.0 mM.

The 98% solution contains 98.00 g of ß-mercaptoethanol per 100.00 g of solution. Using equation 34,

$$\text{vol of 100.00 g solution} = \frac{\text{wt}}{\rho} = \frac{100.00 \text{ g}}{1.114 \text{ g} / \text{mL}} = 89.77 \text{ mL}$$

From equation 8,

$$M = \frac{\text{wt}}{\text{mol wt} \times L} = \frac{98.00 \text{ g}}{78.13 \text{ g} / \text{mol} \times 0.08977 \text{ L}} = 13.97 \text{ M}$$

Let V_1 be the volume of the stock ß-mercaptoethanol needed to prepare the 100.0 mL (V_2) solution, and let C_1 and C_2 be the concentrations of ß-mercaptoethanol in the stock solution and the 100.0 mL solution, respectively. Then, using equation 39,

$$V_1 = \frac{C_2 V_2}{C_1} = \frac{0.05 \text{ M} \times 0.100 \text{ L}}{13.97 \text{ M}} = 0.4 \text{ mL}$$

PROBLEM 13

1.00 g of dry peptone was analyzed and found to contain 50.00 µg of K^+. Calculate (a) the % (w/w) of K^+ in the peptone, (b) the % (w/v) of K^+ in a 5% (w/v) solution of the peptone, and (c) the molarity of K^+ in the 5% solution. The at. wt of K is 39.10.

SOLUTION:

(a) From equation 23,

$$\% \text{ (w/v)} = \frac{\text{g of analyte in sample}}{\text{g of sample}} \times 100\%$$

$$\therefore \% \text{ (w/w) of K}^+ = \frac{5.00 \times 10^{-5} \text{ g of K}^+}{1.00 \text{ g of peptone}} \times 100\%$$

$$= 0.0050\%$$

(b) There are 5.00 g of peptone per 100.0 mL of a 5% solution and 50.00 µg of K$^+$/g of peptone.

$$\therefore \text{µg of K in 5.00 g of peptone} = \frac{50.00 \text{ µg of K}^+}{100.0 \text{ mL of sample}} \times 5.00 \text{ g of peptone}$$

$$= 2.50 \times 10^2 \text{ µg of K}^+$$

Using equation 24,

$$\% \text{ (w/v) of K}^+ = \frac{\text{g of analyte}}{100.0 \text{ mL of sample}} \times 100\%$$

$$= \frac{2.50 \times 10^{-4} \text{ g of K}^+}{100.0 \text{ mL (peptone)}} \times 100\% = 0.000250\%$$

(c) From the result in (b), the 5% solution contains 2.50×10^{-4} g of K$^+$ per 100.0 mL (0.1000 L). Using equation 8,

$$M = \frac{\text{wt}}{\text{mol wt} \times \text{L}} = \frac{2.50 \times 10^{-4} \text{ g}}{39.10 \text{ g / mol} \times 0.100 \text{ L}} = 6.39 \times 10^{-5} \text{ M}$$

PROBLEM 14

A tetraphenyl phosphonium chloride (TPPCl) powder (fw = 342.39) is 96% pure. How many grams are needed to prepare 0.100 L of a 10.0 mM solution?

SOLUTION:

Using equation 10,

$$\text{g of TPPCl needed} = M \times \text{mol wt} \times \text{L} \times \frac{100\%}{\% \text{ purity}}$$

$$= \frac{0.010 \text{ mol / L} \times 342.39 \text{ g / mol} \times 0.100 \text{ L} \times 100\%}{96\%} = 0.36 \text{ g}$$

PROBLEM 15

How many grams of *n*-octyl glucoside are needed to prepare a 5.0% (w/v) solution if
the octyl glucoside is (a) 100% (w/w) pure and (b) 95% (w/w) pure?

SOLUTION:

(a) When 100% purity is assumed, a 5.0% (w/v) solution contains 5.00 g in 100.0 mL
of the solution.

∴ 5.00 g of the 100% pure *n*-octyl glucoside is needed to prepare 100.0 mL of the 5.0%
solution.

Alternatively, the problem can be solved by using equation 23:[4]

$$\% \text{ (w/w)} = \frac{\text{g of analyte}}{\text{g of sample}} \times 100\%$$

$$\therefore \text{g of sample} = \frac{5.00 \text{ g of analyte} \times 100\%}{100\%} = 5.00 \text{ g sample}$$

**To prepare 100.0 mL of a 5.0% (w/v) solution, 5.00 g of the 100% pure *n*-octyl
glucoside is needed.**

(b) The amount of the 95% (w/w) sample containing 5.00 g of pure *n*-octyl glucoside is
similarly calculated using equation 23:[5]

$$\% \text{ (w/w)} = \frac{\text{g of analyte}}{\text{g of sample}} \times 100\%$$

$$\therefore \text{g of sample} = \frac{5.00 \text{ g of analyte}}{95\%} \times 100\% = 5.30 \text{ g of sample}$$

**To prepare 100.0 mL of a 5.0% (w/v) solution, 5.30 g of the 95% *n*-octyl glucoside
is needed.**

PROBLEM 16

The concentration of Na^+ in a standard solution is 1000 ppm (w/v). What is the molarity
of Na^+? The at. wt of Na is 23.00.

SOLUTION:

First, calculate the grams of Na^+ per liter of solution.
By rearranging equation 29,[6]

$$\text{g of analyte} = \frac{\text{ppm of analyte (w/v)} \times \text{mL of sample}}{10^6 \text{ ppm}}$$

$$\therefore \text{g of } Na^+ / L = \frac{1000 \text{ ppm} \times 1000.0 \text{ mL of solution}}{10^6 \text{ ppm}}$$

$$= \frac{1000 \text{ ppm} \times 1000.0 \text{ mL of solution}}{10^6 \text{ ppm}} = 1.00 \text{ g of } Na^+$$

Second, calculate the M of Na^+, using equation 8:

$$M \text{ of } Na^+ = \frac{wt}{at. \ wt \times L} = \frac{1.00 \ g}{23.00 \ g / mol \times L} = 0.04 \ mol / L \ \text{ or } 0.04 \ M$$

The standard solution is 0.04 M in Na^+.

PROBLEM 17

Trace Pb^{2+} in $MgSO_4$ is 500 ppm (w/w). Calculate the molarity of Pb^{2+} in a 20% (w/v) solution of the $MgSO_4$. The at. wt of Pb is 207.20.

SOLUTION:

First, calculate the grams of Pb^{2+} in the 20.00 g of $MgSO_4$ needed to prepare a 20% (w/v) solution. From equation 23 (rearranged),[7]

$$g \text{ of analyte} = \frac{ppm \ (w/w) \text{ of analyte} \times g \text{ of sample}}{10^6 \ ppm \text{ of sample}}$$

$$\therefore g \text{ of } Pb^{2+} / 20.00 \ g \text{ of } MgSO_4 = \frac{500 \ ppm \text{ of } Pb^{2+} \times 20.00 \ g \text{ of } Pb^{2+}}{10^6 \ ppm} = 0.010 \ g$$

Second, using equation 8, calculate the molarity of Pb^{2+}.

$$M \text{ of } Pb^{2+} = \frac{wt}{mol \ wt \times L} = \frac{0.010 \ g}{207.20 \ g / mol \times 0.100 \ L} = 4.8 \times 10^{-4} \ M \ \text{ or } 480 \ \mu M$$

PROBLEM 18

Prepare 0.500 L of a 0.200 N solution of (a) NaOH and (b) $Ca(OH)_2$. The fw of NaOH and $Ca(OH)_2$ are 40.0 and 74.1, respectively.

SOLUTION:

(a) NaOH contains 1 OH^-/molecule. Therefore, the number of equiv/mole = n = 1. Using equation 3,

$$equiv \ wt \text{ of } NaOH = \frac{mol \ wt}{n} = \frac{40.0 \ g / mol}{1.0 \ equiv / mol} = 40.0 \ g / equiv$$

Using equation 18,

$$g \text{ of } NaOH \text{ need } = N \times equiv \ wt \times L$$

$$= 0.200 \ equiv / L \times 40.0 \ g / equiv \times 0.500 \ L = 4.00 \ g$$

Dissolve 4.00 g of NaOH in 250.0 mL of H_2O. Cool, then adjust volume to 500.0 mL with additional H_2O. See section 1.5 for method.

(b) Similarly, $Ca(OH)_2$ contains 2 OH^-/molecule.

$$\therefore n = 2$$

$$equiv \ wt \text{ of } Ca(OH)_2 = \frac{74.1 \ g / mol}{2.0 \ equiv / mol} = 37.05 \ g / equiv$$

g $Ca(OH)_2$ needed = 0.200 equiv / L × 37.05 g / equiv × 0.500 L = 3.70 g

Dissolve 3.70 g of $Ca(OH)_2$ in 250.0 mL of H_2O. Adjust volume to 500.0 mL with additional H_2O. See section 1.5 for method.

PROBLEM 19

Calculate the number of (a) base equivalent in 5.00 g of NaOH, (b) acid equivalents in 1.0 mL of a 0.10 $(NH_4)_2SO_4$ solution, and (c) electron equivalents in 50.0 mL of a 10.0 mM solution of a compound containing 1 mole of transferable electrons per mole. The mol wt of NaOH is 40.0.

SOLUTION:

(a) NaOH contains 1.0 base equiv/mole (i.e., 1.0 OH⁻/molecule); $n = 1$.
Using equation 3,

$$\text{equiv wt of NaOH} = \frac{\text{mol wt}}{n} = \frac{40.0\ g\ /\ mol}{1.0\ \text{quiv}\ /\ mol} = 40.0\ g\ /\ mol$$

Using equation 4,

$$\text{No. of equiv} = \frac{wt}{\text{equiv wt}} = \frac{5.00\ g}{40.0\ g\ /\ equiv} = 0.12\ \text{equiv}$$

(b) A 0.10 N solution contains 0.10 equiv/L; i.e., 1000.0 mL of 0.10 $(NH_4)_2SO_4$ contains 0.10 equiv.

$$\therefore 1.0\ \text{mL contains}\ \frac{0.10\ \text{equiv}}{1000.0\ \text{mL}} \times 1.0\ \text{mL} = 1.0 \times 10^{-4}\ \text{equiv}$$

(c) The compound contains 1.0 mole of transferable electrons/mol.

$$\therefore \text{The number of equiv / mol} = n = 1;\ M = 0.0100$$

Using equation 21,

$$N = nM = 1.0\ \text{equiv / mol} \times 0.0100\ \text{mol / L} = 0.0100\ \text{equiv / L}$$

This means that 1000.0 mL contains 0.0100 electron equiv

$$\therefore 50.0\ \text{mL contains}\ \frac{0.0100\ \text{equiv}}{1000.0\ \text{mL}} \times 50.0\ \text{mL} = 5.0 \times 10^{-4}\ \text{equiv}$$

PROBLEM 20

The concentration of HCl in a commercial concentrated HCl solution is given as 37.0% (w/w). Calculate (a) the molarity and (b) the normality of HCl in the solution. (c) Prepare 100.0 mL of a 2.0 N solution of the HCl. The fw and sp gr of the HCl are 36.50 and 1.19, respectively.

SOLUTION:

(a) First, calculate the volume of 100.00 g of the HCl solution. Using equation 34,

$$\text{vol of 100.0 g of solution} = \frac{\text{g of solution}}{\rho \text{ of solution}} = \frac{100.00 \text{ g}}{1.19 \text{ g / mL}} = 84.0 \text{ mL}$$

Second, calculate the molarity using equation 8; 100.00 g of the solution contains 37.00 g of pure HCl in a volume of 0.0840 L.

$$\therefore \text{M of HCl} = \frac{\text{wt}}{\text{mol wt} \times \text{L}} = \frac{37.00 \text{ g}}{36.50 \text{ g / mol} \times 0.0840 \text{ L}} = 12.1 \text{ M}$$

(b) HCl contains 1 H^+ per molecule. The number of equiv/mol = n = 1. Using equation 21,

$$\text{N} = n\text{M} = 1 \text{ equiv / mol} \times 12.06 \text{ mol / L} = 12.06 \text{ equiv / L or 12.1 N}$$

(c) Calculate the volume of concentrated HCl needed using equation 39: Let V_1 be the volume of concentrated HCl needed to prepare 100.0 mL (V_2) of the 2.0 N solution; C_1 and C_2 are the concentrations of HCl in the concentrated and 2 N solutions, respectively. From equation 39,

$$C_1 V_1 = C_2 V_2$$

$$\therefore V_1 = \frac{C_2 V_2}{C_1} = \frac{2.0 \text{ N} \times 100.0 \text{ mL}}{12.06 \text{ N}} = 17.0 \text{ mL}$$

Add 17.0 mL of concentrated HCl slowly to 83.0 mL of H_2O to prepare 100.0 mL of the 2.0 N solution. See section 1.5 for method.

PROBLEM 21

Calculate the normality of commercial concentrated H_2SO_4 and H_3PO_4. The % concentrations, fw, and sp gr, respectively, are: H_2SO_4, 95.0% (w/w), 98.1, 1.84; H_3PO_4, 85% (w/w), 98.0, 1.71.

SOLUTION:

Calculate the molarity of H_2SO_4 and H_3PO_4 following the example in problem 19a. These values are: H_2SO_4, 17.8 M; H_3PO_4, 14.7 M.

H_2SO_4 contains 2 H^+/molecule; the number of equiv/mol = n = 2. H_3PO_4 contains 3 ionizable H^+/molecule; the number of equiv/mol = n = 3. The normality is calculated using equation 21,

$$\text{N} = n\text{M}$$

For H_2SO_4,

$$\text{N} = 2 \text{ equiv / mol} \times 17.8 \text{ mol / L} = 35.6 \text{ equiv / L or 35.6 N}$$

For H_3PO_4,

$$\text{N} = 3 \text{ equiv / mol} \times 14.7 \text{ mol / L} = 44.1 \text{ equiv / L or 44.1 N}$$

PROBLEM 22

Prepare 100.0 mL of a 2.0 N solution of (a) H_2SO_4 and (b) H_3PO_4, starting with the concentrated acids.

SOLUTION:

Calculate the normality of concentrated H_2SO_4 and H_3PO_4 (problem 20) or obtain it from Appendix D, Table D1. These values are 35.6 N for H_2SO_4 and 44.1 N for H_3PO_4.

(a) Let V_1 be the volume of concentrated H_2SO_4 needed to prepare 100.0 mL (V_2) of the 2.0 N solution; C_1 and C_2 are the concentrations of H_2SO_4 in the concentrated and 2.0 N solutions, respectively. From equation 39,

$$C_1V_1 = C_2V_2$$

$$\therefore V_1 = \frac{C_2V_2}{C_1} = \frac{2.0\ \text{N} \times 100.0\ \text{mL}}{35.6\ \text{N}} = 5.6\ \text{mL}$$

Add 5.6 mL of concentrated H_2SO_4 slowly to 94.4 mL of H_2O to prepare 100.0 mL of the 2.0 N solution. See section 1.5 for method.

(b) As an exercise, calculate the volume of H_3PO_4 needed to prepare the 100.0 mL of the 2.0 N solution (answer = 4.5 mL).

PROBLEM 23

A 75.0 μL aliquot of a solution of EcoR1 (100.0 units/mL) is added to 0.50 mL of buffer. Calculate (a) the dilution factor (DF) and (b) the final concentration of EcoR1 in the buffer. (c) Prepare 1.50 mL of a 15.0 units/mL solution starting with the 100.0 units/mL stock.

SOLUTION:

(a) From equation 38,

$$DF = \frac{V_2}{V_1} = \frac{0.50\ \text{mL} + 0.0750\ \text{mL}}{0.0750\ \text{mL}} = 7.7$$

(b) Let V_1 be the volume of the 100.0 units/mL (C_1) solution added to the 0.50 mL (V_2) buffer, and let C_2 be the final concentration of EcoR1. Then, using equation 39, the final concentration is:

$$C_2 = \frac{C_1V_1}{V_2} = \frac{100.0\ (\text{units / mL}) \times 0.0750\ \text{mL}}{0.575\ \text{mL}} = 13.0\ \text{units / mL}$$

Alternatively, the final concentration can be calculated as follows:

$$C_2 = \frac{C_1}{DF} = \frac{100.0\ \text{units / mL}}{7.70} = 13.0\ \text{units / mL}$$

(c) Let V_1 be the volume of the stock solution needed to prepare 1.50 mL (V_2) of the 15.0 units/mL solution. C_1 and C_2 are the concentrations of EcoR1 in the stock and the 15.0 units/mL solutions, respectively. From equation 39,

$$V_1 = \frac{C_2 V_2}{C_1} = \frac{15.0 \text{ units / mL} \times 1.50 \text{ mL}}{100.0 \text{ units / mL}} = 0.225 \text{ mL}$$

Add 225 µL of the stock solution to 1.275 mL of diluent to prepare 1.50 mL of the 15.0 units/mL solution.

PROBLEM 24

How would you prepare 500.0 mL of 0.45M Copper(II)chloride dihydrate from the solute?

SOLUTION:

Calculation

$CuCl_2 \cdot 2H_2O$
Cu = 63.55
2Cl = 70.90
2O = 32.00
4H = 4.04
Total = 170.49g/mol

$$\text{g } CuCl_2 \cdot 2H_2O = M \times MW \times L$$
$$= (0.45 \text{ M}) \times (170.49 \text{ g / mol}) \times (0.500 \text{ L})$$
$$= 38.36$$

Method

1. Weigh out 38.36 g of $CuCl_2 \cdot 2H_2O$.
2. Add about $\dfrac{500.0}{2}$ or 250.0 mL of water into a volumetric flask or graduated cylinder.
3. Add the 38.36 g of $CuCl_2.2H_2O$ to the water and mix until it dissolves completely.
4. Adjust the volume to 500.0 mL by adding more water.
5. Mix the solution thoroughly.

PROBLEM 25

How would you prepare 8.0 liters of 0.50 M lithium sulfate from the solute?

SOLUTION:
Calculation

Li_2SO_4
2Li = 2(6.94)
1S = 1(32.1)
4O = 4(16.0)
Total = 109.98 g/mol

$$\text{g } Li_2SO_4 = M \times MW \times L$$
$$= (0.50 \text{ M}) \times (109.98 \text{ g / mol}) \times (8.0 \text{ L})$$
$$= 439.92 \text{ g}$$

Method
1. Weigh out 439.92 g of Li_2SO_4.

2. Add about $\dfrac{8.0}{2}$ or 4.0 L of water into a container large enough to contain 8 L.

3. Add the 439.92 g of Li_2SO_4 to the 4.0 L of water and mix until it dissolves completely.

4. Adjust the volume to 8.0 L by adding more water.

5. Mix the solution thoroughly.

PROBLEM 26

How would you prepare 865.0 mL of a 3.3 M solution of sulfuric acid, starting from an 8.9 M stock solution?

SOLUTION:
From equation 40, $M_1V_1 = M_2V_2$

Calculation

$$M_1 = 8.9 \text{ M}$$
$$V_1 = \text{question}$$
$$M_2 = 3.3 \text{ M}$$
$$V_2 = 865.0 \text{ mL}$$
$$M_1V_1 = M_2V_2$$

$$V_1 = \frac{M_2V_2}{M_1} = \frac{3.3 \text{ M} \times 865.0 \text{ mL}}{8.9 \text{ M}} = 320.7 \text{ mL}$$

Method
1. Measure out 320.7 mL of the 8.9 M sulfuric acid.
2. Add about (865.0 - 320.7)/2 or 272.2 mL of water in a volumetric flask or graduated cylinder.
3. Add the 320.7 mL of 8.9 M sulfuric acid **slowly** and mix.
4. Adjust the volume to 865.0 mL by adding more water.
5. Mix thoroughly.

PROBLEM 27

Starting from a 2.5 M acetic acid solution, how would you prepare a 112.0 mL solution with a final concentration of 0.75M?

SOLUTION:
From equation 40, $M_1V_1 = M_2V_2$

Calculation

$$M_1 = 2.5 \text{ M}$$
$$V_1 = \text{question}$$
$$M_2 = 0.75 \text{ M}$$
$$V_2 = 112.0 \text{ mL}$$
$$M_1V_1 = M_2V_2$$

$$V_1 = \frac{M_2V_2}{M_1} = \frac{0.75 \text{ M} \times 112.0 \text{ mL}}{2.5 \text{ M}} = 33.6 \text{ mL}$$

Method

1. Measure out 33.6 mL of 2.5 M acetic acid.
2. Add about(112.0 - 33.6)/2 or 39.2 ml of water in a volumetric flask or graduated cylinder.
3. Add the 33.6 mL of 2.5 M acetic acid and mix.
4. Adjust the volume to 112.0 mL by adding more water.
5. Mix thoroughly.

---

[1] For the units in these equations to cancel out, the %, ppm, or ppb units in the equations must be substituted with their gram and/or volume equivalents. For example, when dealing with % (w/w), replace (% analyte) with (g of analyte/100.00 g of sample), and the % to the right of the equation with (g of sample/100.00 g of sample); and when dealing with % (w/v), replace (% analyte) with (g of analyte/100.0 mL of sample), and the % to the right with (mL of sample/100.0 mL of sample). See problems 14 and 15 in section 1.6.

[2] If the diluent volume is equal to the target final volume prior to dissolving the stock reagent, the final volume of the solution will exceed the target final volume, resulting in a lower-than-expected concentration. This is because the solute molecules occupy a volume in the solution. See section 1.3.I.

[3] See footnote 2.

[4] See footnote 1.

[5] See footnote 1.

[6] See footnote 1.

[7] See footnote 1.

Chapter 2

Calculations Involving Chemical Reaction and Stoichiometry

2.1. Principles of Chemical Stoichiometry

Stoichiometry is the study of quantitative relationships between substances undergoing chemical changes in both chemical and biological systems. It is an important concept in both theoretical and applied aspects of chemical reactions. For example, calculations for the composition of matter and the amount of reactants needed to produce certain amounts of products rely on the principles of stoichiometry. A wide range of materials are usually covered in this area, but we will limit our discussions to just a few of them, including chemical reactions, equations, limiting reactant, percent yield, rates, molar mass, percent composition and formulae. An adequate number of problems are solved to illustrate the principles from different perspectives.

2.2. Molar Mass and Percent Composition of Compounds

A. Molar Mass

One mole of a compound contains Avogadro's number of formula units of that compound. If a compound is composed of molecules (for example H_2O, CO_2, NH_3, etc.) and the formula of the compound is known, its molar mass may be determined by adding together the molar masses of all the atoms in the formula. If more than one atom of an element is present, the mass of that element is added as many times as it is used. For example the mass of 1 mole of water, formula H_2O, can be found by summing the masses of hydrogen and oxygen present:

$$2\,H = 2 \times 1.01\ g = 2.02\ g$$

$$1\,O = 1 \times 16.00\ g = 16.00\ g$$

$$Sum = 18.02\ g\,/\,mol$$

This number represents the mass of 1 mole of water molecule. Therefore the **molar mass** of a substance is the mass in grams of one mole of the compound. Molar mass and molecular weight (section 1.1F) describe the same quantity and are, therefore, equivalent terms. For ionic compounds, the sum of the atomic masses of the elements present in one formula unit is known as the formula mass or formula weight (section 1.1G).

B. Percent Composition

As described in chapter 1 (section 1.3), percent means parts per one hundred parts. The percent composition of a compound is the mass percent of each element in the compound. The molar mass represent the total mass, or 100%, of the compound [see percent (w/w), section 1.3A)]. If the chemical formula of a compound is known, its percent composition can be determined as follows:

1. Calculate the molar mass.
2. For each element in the compound, divide its mass by the molar mass of the compound and multiply by 100. Thus,

$$\% \text{ composition for each element} = \frac{\text{mass of the element}}{\text{molar mass of compound}} \times 100\% \qquad (41)$$

Examples: See problems 28–32.

2.3. Calculating the Empirical and Molecular Formulae of Compounds Using Molar Mass and Percent Composition Data

A. Empirical versus Molecular Formulae

There are different kinds of chemical formulas. Each of them conveys a certain kind of information. The empirical formula is the formula that gives the smallest whole-number ratio of the different atoms in a compound, or that uses the smallest set of whole-number subscripts to specify the relative number of atoms of each element in a formula unit. The empirical formula is the simplest formula.

The molecular formula (see section 1.1E) is the true formula. It states the actual ratio of each kind of atom found in one molecule of a compound. The formula C_2H_4 is a molecular formula for ethylene. It has two atoms of carbon and four atoms of hydrogen. However, its simplest formula is CH_2 since the carbon to hydrogen ratio is 1 to 2. This simplest formula is not unique to C_2H_4. A substance whose empirical formula is CH_2 could have as a molecular formula CH_2, C_2H_4, C_3H_6, C_4H_8, C_5H_{10}, etc.

B. Calculating Empirical Formula from Percent Composition and Molar Mass Data

The empirical formula of a compound can be calculated if the molar mass, percent composition and other pertinent data are known. The calculation involves the following steps:

Step 1: Assume you have 100 grams of the substance, if you are given percent composition.

Step 2: Multiply the given mass (in grams) of each element by the factor (1 mol/1 molar mass of the element) to convert grams to moles. This conversion gives the number of moles of atoms of each element in the given mass. Keep at least one extra significant figure in this and all subsequent steps. At this point these numbers will usually not be whole numbers.

Step 3: Divide each of the values obtained in Step 2 by the smallest of them. If the numbers obtained are whole numbers, or are within ± 0.05 of an integer, round off and stop. You have the answer. Use them as subscripts in writing the empirical formula. If not perform step 4.

Step 4: Multiply the values obtained in Step 3 by the smallest number that will convert them to whole numbers. Use these whole numbers as the subscripts in the empirical formula.

Examples: See problems 33–37.

C. Calculating Molecular Formula from Empirical Formula and Molar Mass Data

The rationale for calculating molecular formula from empirical formula is as follows. If the empirical formula of a compound of carbon and hydrogen is CH_2 then the molecular formula can be expressed as $(CH_2)_\eta$, where $\eta = 1, 2, 3, 4, \ldots$ Thus, the molecular formula could be $CH_2, C_2H_4, C_3H_6, C_4H_8, C_5H_{10}, \ldots\ldots$ The quantity, η, is the number of empirical formula units and its actual value is calculated using the following equation:

$$\eta = \frac{\textbf{molar mass of the molecule}}{\textbf{mass of empirical formula}} \tag{42}$$

Examples: See problems 38–42.

2.4. Calculations Involving Chemical Reactions

A. Chemical Reactions

A chemical reaction involves a chemical change in which the atoms in one or more chemical substances are reorganized to produce different substances. For example, when propane (C_3H_8) in natural gas combines with oxygen (O_2) in the air and burns, atoms in both the propane and oxygen are reorganized to produce carbon dioxide (CO_2) and water (H_2O).

B. Chemical Equations

A chemical reaction is traditionally represented as a chemical equation using chemical symbols. In a chemical equation, an arrow sign is written in the middle to separate reactants (C_3H_8 and O_2 in the above example) that are written to the left, from products (CO_2 and H_2O) that are written to the right. Below is the chemical equation for the above reaction:

$$C_3H_8 + O_2 \rightarrow CO_2 + H_2O$$

It is often useful to indicate whether the reactants and products are solids (s), liquids (l) or gases (g). This is done by writing the symbols s, l or g in parentheses to the right of each reactant and product. For example, the above reaction becomes:

$$C_3H_8\,(g) + O_2\,(g) \rightarrow CO_2\,(g) + H_2O(l)$$

Also, a vast majority of the reactions encountered in chemical and biological systems occur when two substances dissolved in water are mixed. These solutions are known as aqueous solutions (from the Latin word aqua for water). The symbol aq is used to identify such solutions. Thus, solid sugar is written $C_{12}H_{22}O_{11}$ (s) and an aqueous solution of sugar is written as $C_{12}H_{22}O_{11}$ (aq).

C. Balancing Chemical Equations

A close inspection of the above equation shows that it is not balanced; i.e. the total number of atoms of each element in the reactants and the products are not equal. Thus, there are 3 atoms of C in the reactants and only one atom in the products. The equation is balanced by applying the principles of conservation, which states that atoms are conserved in a chemical reaction. In balancing a chemical reaction, the identities of the products and reactants must not be changed since they are experimentally determined. The formulae of the compounds must not be changed either. This means that the equation cannot be balanced by simply changing the subscripts in a formula or by adding or subtracting atoms from the formula. Chemical equations are usually balanced by inspection or trial and error during which **coefficients** are introduced as multiplication factors to equalize the number of atoms between reactants and products. It is best to start with the most complicated molecule. As an example, let us now balance the above equation.

$$C_3H_8(g) + O_2(g) \rightarrow CO_2(g) + H_2O(l)$$

The most complicated moleculae is $C_3H_8(g)$. We start out by balancing the product that contains the atoms in $C_3H_8(g)$. $C_3H_8(g)$ contains three C atoms. To balance for C, we place the coefficient 3 before the product $CO_2(g)$ to obtain 3 C atoms in the product. Since $C_3H_8(g)$ contains eight H atoms, whereas, there are only 2 H atoms in the product $H_2O(l)$, we write the coefficient 4 before the $H_2O(l)$ to obtain 8 H atoms for the products. There are now 10 O atoms in the products but only two in the reactants. We apply the same logic to balance for O by writing the coefficient 5 before the reactant, $O_2(g)$. The balanced equation is now:

$$C_3H_8(g) + 5O_2(g) \rightarrow 3CO_2(g) + 4H_2O(l)$$

Finally, we check for balance.

Atoms of reactants	Atoms of products
3 C	3 C
8 H	8 H
10 O	10 O

Thus the equation is balanced.

D. Limiting Reactant

For several chemical reactions or processes, the quantities of the reactants used are such that the amount of one reactant is in excess of the amount of the other reactants. The amount of the product(s) formed will, thus, depend on the reactant that is **not** in excess. This brings us to the concept of the limiting reactant (or limiting reagent). The limiting reactant is the one that is not in excess. It gets consumed first and, therefore, limits the amounts of products that can be formed. In any stoichiometry calculation involving a chemical reaction, it is essential to determine which reactant is limiting in order to calculate correctly the amounts of products formed. As an example, consider a case in which a solution of 1.0 mole of sodium hydroxide is mixed with a solution of 1.5 moles hydrochloric acid. The following reaction occurs:

$$NaOH + HCl \rightarrow NaCl + H_2O$$

Is this a stoichiometric mixture of reactants, or will one of the reactants be consumed before the other? According to the equation, it is possible to obtain 1.0 mol of NaCl from 1.0 mol of NaOH, and 1.5 mol of NaCl from 1.5 mol of HCl. However, we cannot have two different yields of NaCl from this reaction. When 1.0 mol of NaOH and 1.5 mol of HCl are mixed, there is insufficient NaOH to completely react with all of the HCl. So, HCl is the reactant in excess and NaOH is the limiting reactant. Since the amount of NaCl formed is dependent on the limiting reactant, only 1.0 mol of NaCl will be formed, 0.5 mol of HCl remains unreacted. In solving "limiting reactant" problems, we first identify which is the limiting reactant. Then, we calculate the amount of product formed based on the amount of the limiting reactant available.

Examples: See problems 43–47.

E. Percent Yield

In chemical reactions, two kinds of yields are commonly considered; theoretical yield, and actual or experimental yield. Theoretical yield is a calculated quantity based on the principles of stoichiometry. In other words, the theoretical yield of a given product is the maximum yield that can be obtained if the reactants reacted completely and gave no side reactions. In actual practice, factors such as impure reactants, incomplete reactions, and side reactions can cause the actual yield to be less than the calculated one. If we know the actual yield obtained experimentally and the theoretical yield calculated by stoichiometry, we can find the percentage yield as:

$$\% \text{ yield} = \frac{\text{actual yield}}{\text{theoretical yield}} \times 100\% \qquad (43)$$

Examples: See problems 48–51.

F. Reaction Rate and Order

In this section we begin to examine the factors that control the rate or speed (i.e. how fast or slow) a chemical reaction takes place. For ionic reactions, the speed is determined by how rapidly we can mix the chemicals. Other reactions, such as the digestion of food, takes place more slowly. These different rates exist because of chemical differences among the reacting substances.

For any given reaction, one of the most important controlling influences is the concentrations of the reactants. Generally if we follow a chemical reaction over a period of time, we find that its rate gradually decreases as the reactants are consumed.

The rate or speed of a process is defined as the change in a given quantity over a specific period of time. For a chemical reaction, the quantity that changes is the amount or concentration of the reactant or product. For the general reaction shown in equation 44 below:

$$A \rightarrow B \qquad (44)$$

the rate of the reaction is defined as the change in concentration of the reactant (A) or the product (B) per unit time:

$$\text{Rate} = \frac{\text{concentration of A at time } t_2 - \text{concentration of A at time } t_1}{t_2 - t_1} = \frac{\Delta[A]}{\Delta t} \qquad (45)$$

The square brackets indicate concentration in mol/L. The symbol, Δ, indicates a change in a given quantity. Note that a change can be positive (increase) or negative (decrease). As such, we can have a positive or negative reaction rate. Mathematical expressions for both positive and negative rates can be written for the reaction shown below.

$$1\,A + 2\,B \rightarrow 1\,C + 3\,D \tag{46}$$

The rates are:

$$-\frac{\Delta[A]}{\Delta t},\ \text{or} -\frac{\Delta[B]}{\Delta t},\ \text{or} +\frac{\Delta[C]}{\Delta t},\ \text{or} +\frac{\Delta[D]}{\Delta t}$$

For the general reaction shown in equation 44 above, the rate can be written as

$$\text{rate} \propto [A]^x$$

where the exponent x, is called the **order of the reaction**. When x = 1, the reaction is first-order. When x = 2 or 3, the reaction is second and third order, respectively. It is possible to have higher orders as well as fractional order of reactions. Consider the slightly more complex reaction below,

$$A + B \rightarrow \text{products} \tag{47}$$

the rate usually depends on the concentrations of both A and B. Normally, increasing the concentration of either A or B will increase the reaction rate, and the rate is then proportional to the product of the concentrations of A and B, each raised to some power.

$$\text{rate} \propto [A]^x[B]^y \tag{48}$$

For the rate shown in 48, we say that the order of the reaction with respect to A is x and the order with respect to B is y. We can also describe the **overall order** of the reaction as the sum of the exponents of the concentration terms, which in this case is (x+y). When one of the exponents is zero, this means that the rate of the reaction does not depend on the concentration of that particular substance. For instance if the exponent y in equation 48 were zero, the equation would become:

$$\text{rate} \propto [A]^x[B]^0$$

which reduces to:

$$\text{rate} \propto [A]^x$$

The proportionality in equation 48 can be converted to an equality by introducing a proportionality constant, k, which is called the **rate constant**. The resulting equation is called the **rate law** for the reaction. Thus,

$$\text{rate} \propto [A]^x[B]^y\ \text{becomes}$$

$$\text{rate k}[A]^x[B]^y \tag{49}$$

Take the decomposition of nitrogen dioxide for an example.

$$2NO_2(g) \rightarrow 2NO(g) + O_2(g)$$

This reaction can occur as written and can also occur in the reverse direction as NO and O_2 accumulate to reform NO_2:

$$2NO(g) + O_2(g) \rightarrow 2NO_2(g)$$

Thus when gaseous NO_2 is placed in an empty container, initially the dominant reaction is

$$2NO_2(g) \rightarrow 2NO(g) + O_2(g)$$

and the change in the concentration of NO_2 depends only on the forward reaction. After a period of time, however, enough products accumulate and the reverse reaction becomes important. The change in concentration of NO_2 at this time depends on the difference in the rates of the forward and the reverse reactions. This complication can be avoided if we study the rate of the reaction under the conditions where the contribution of the reverse reaction is considered negligible. This means that we must study the reaction as soon as the reactants are mixed, before the products have had enough time to build up to significant levels.

For the decomposition of $NO_2(g)$ when the reverse reaction is neglected:

$$Rate = k[NO_2]^2$$

Once again k is called the **rate constant**, the order of the reaction is 2 and the reaction is said to be second order. The values of both the rate constant and order must be experimentally determined. It is to be noted that there is no direct relationship between the coefficients in the balanced chemical equation and the order of the reaction. However the rate at which a reactant decreases is proportional to the coefficient of that reactant, and the rate at which a product increases is proportional to the coefficient of that product.

Examples: See problems 52–54.

DEFINITIONS

For the general reaction shown in equation 44 above, the rate law is written as

$$Rate = k[A]^x$$

$$\text{zero order: } Rate = k[A]^0$$

$$\text{first order: } Rate = k[A]^1$$

$$\text{second order: } Rate = k[A]^2$$

$$\text{third order: } Rate = k[A]^3$$

Examples: See problems 56, 57, and 58

2.5. Practical Examples

PROBLEM 28

Determine the percent composition of all the elements in copper(II)sulfate.

SOLUTION:

First calculate the molar mass for $CuSO_4$.

$$1Cu = 1(63.5) = 63.5$$
$$1S = 1(32.1) = 32.1$$
$$4O = 4(16.0) = 64.0$$
$$Total = 159.6 \text{ g} = \text{molar mass of } CuSO_4$$

Then calculate the percentage of each element in $CuSO_4$.

$$\%Cu = \frac{g\ Cu}{g\ total} \times 100\% = \frac{63.5\ g}{159.6\ g} \times 100\% = 39.8\%$$

$$\%S = \frac{g\ S}{g\ total} \times 100\% = \frac{32.1\ g}{159.6\ g} \times 100\% = 20.1\%$$

$$\%O = \frac{g\ O}{g\ total} \times 100\% = \frac{64.0\ g}{159.6\ g} \times 100\% = 40.1\%$$

PROBLEM 29

Determine the percent composition of all the elements in sodium sulfate.

SOLUTION:

The molecular formula of sodium sulfate is Na_2SO_4.
First calculate the molar mass.

$$2Na = 2(23.0) = 46.0$$
$$1S = 1(32.1) = 32.1$$
$$4O = 4(16.0) = 64.0$$
$$Total = 142.1 \text{ g} = \text{molar mass of } Na_2SO_4.$$

Then calculate the percentage of each element in Na_2SO_4.

$$\%Na = \frac{g\ Na}{g\ total} \times 100\% = \frac{46.0\ g}{142.1\ g} \times 100\% = 32.4\%$$

$$\%S = \frac{g\ S}{g\ total} \times 100\% = \frac{32.1\ g}{142.1\ g} \times 100\% = 22.6\%$$

$$\%O = \frac{g\ O}{g\ total} \times 100\% = \frac{64.0\ g}{142.1\ g} \times 100\% = 45.0\%$$

PROBLEM 30

When heated in air, 1.60 g of zinc combined with 0.43 g of oxygen forms zinc oxide (ZnO). Calculate the percentage composition of the compound formed.

SOLUTION:
First, calculate the total mass of the compound formed.

$$1.60 \text{ g Zn} + 0.43 \text{ g O}_2 = 2.03 \text{ g} = \text{total mass of product}$$

Then calculate the % composition based on this mass.

$$\%Zn = \frac{\text{g Zn}}{\text{g total}} \times 100\% = \frac{1.6 \text{ g}}{2.01 \text{ g}} \times 100\% = 78.8\%$$

$$\%O = \frac{\text{g O}}{\text{g total}} \times 100\% = \frac{0.43 \text{ g}}{2.03 \text{ g}} \times 100\% = 21.2\%$$

PROBLEM 31
Trinitrotoluene, an explosive commonly known as TNT, has the formula $C_7H_5N_3O_6$. What is the percent composition of TNT?

SOLUTION:
First calculate the molar masses for the elements in $C_7H_5N_3O_6$.
$$7C = 7(12.01) = 84.07$$
$$5H = 5(1.01) = 5.05$$
$$3N = 3(14.0) = 42.0$$
$$6O = 6(16.0) = 96.0$$
$$\text{Total} = 227.12 \text{ g} = \text{molar mass of } C_7H_5N_3O_6$$
Then calculate the percentage of each element in $C_7H_5N_3O_6$.

$$\%C = \frac{\text{g C}}{\text{g total}} \times 100\% = \frac{84.07 \text{ g}}{227.12 \text{ g}} \times 100\% = 37.0\%$$

$$\%H = \frac{\text{g H}}{\text{g total}} \times 100\% = \frac{5.05 \text{ g}}{227.12 \text{ g}} \times 100\% = 2.22\%$$

$$\%N = \frac{\text{g N}}{\text{g total}} \times 100\% = \frac{42.0 \text{ g}}{227.12 \text{ g}} \times 100\% = 18.5\%$$

$$\%O = \frac{\text{g O}}{\text{g total}} \times 100\% = \frac{96.0 \text{ g}}{227.12 \text{ g}} \times 100\% = 42.3\%$$

PROBLEM 32
Phenol has the molecular formula, C_6H_5OH. Calculate its percent composition.

SOLUTION:
There are six carbon, six hydrogen, and one oxygen atoms in one molecule of the compound.

Next, calculate the molar masses.
$$6C = 6(12.01) = 72.06$$
$$6H = 6(1.01) = 6.06$$
$$1O = 1(16.0) = 16.0$$

$$Total = 94.12 \text{ g} = \text{molar mass of phenol}$$

Then, calculate the percentage of each element in the compound.

$$\%C = \frac{g\ C}{g\ total} \times 100\% = \frac{72.06\ g}{94.12\ g} \times 100\% = 76.6\%$$

$$\%H = \frac{g\ H}{g\ total} \times 100\% = \frac{6.06\ g}{94.12\ g} \times 100\% = 6.4\%$$

$$\%O = \frac{g\ O}{g\ total} \times 100\% = \frac{16.0\ g}{94.12\ g} \times 100\% = 17.0\%$$

PROBLEM 33

The analysis of a substance showed that it contained 56.58% potassium, 8.68% carbon, and 34.73% oxygen. Calculate the empirical formula for this substance.

SOLUTION:

Apply the steps given in section 2.3B.

Step 1: 100 g total of the substance is assumed. Therefore, the % of each element is equal to its gram weight per 100 g of the substance.

Step 2: Multiply by the proper (1 mol/molar mass) factor to obtain the relative number of moles of each element per mole of the substance.

$$K: 56.58 \text{ g K} \times \frac{1 \text{ mol K atoms}}{39.1 \text{ g K}} = 1.45 \text{ mol K atoms}$$

$$C: 8.68 \text{ g C} \times \frac{1 \text{ mol K atoms}}{12.01 \text{ g C}} = 0.723 \text{ mol C atoms}$$

$$O: 34.73 \text{ g O} \times \frac{1 \text{ mol O atoms}}{16.0 \text{ g O}} = 2.17 \text{ mol O atoms}$$

Step 3: Divide each relative mole by the smallest value.

$$K = \frac{1.45 \text{ mol}}{0.723 \text{ mol}} = 2.00 \text{ mol}$$

$$C = \frac{0.723 \text{ mol}}{0.723 \text{ mol}} = 1.00 \text{ mol}$$

$$O = \frac{2.17 \text{ mol}}{0.723 \text{ mol}} = 3.00 \text{ mol}$$

Therefore, the simplest ratio of K:C:O is 2:1:3.

The empirical formula = **K_2CO_3**.

PROBLEM 34

A sulfide of iron was formed by combining 2.233 g of iron, with 1.926 g of sulfur. What is the empirical formula of the compound?

SOLUTION:

Steps 1 and 2: Since the grams of each element are given, we can use them directly in the calculations. Calculate the relative number of moles of each element by multiplying grams of each element by the proper mol/molar mass factor.

$$\text{Fe: } 2.233 \text{ g Fe} \times \frac{1 \text{ mol Fe atoms}}{55.8 \text{ g Fe}} = 0.0400 \text{ mol Fe atoms}$$

$$\text{S: } 1.926 \text{ g S} \times \frac{1 \text{ mol S atoms}}{32.1 \text{ g S}} = 0.0600 \text{ mol S atoms}$$

Step 3: Divide each number of moles by the smaller of the two numbers.

$$\text{Fe} = \frac{0.0400 \text{ mol}}{0.0400 \text{ mol}} = 1.00$$

$$\text{S} = \frac{0.0600 \text{ mol}}{0.0400 \text{ mol}} = 1.50$$

Step 4: Multiply each value by the smallest factor that gives whole number of atoms. This factor is 2, and multiplication results in a ratio of 2.00 atoms of Fe to 3.00 atoms of S. The empirical formula = **Fe_2S_3**.

PROBLEM 35

A 0.1000 g sample of an alcohol, known to contain carbon, hydrogen and oxygen only, was burned completely to form carbon dioxide (CO_2) and water (H_2O). These products were trapped separately and weighed. 0.1910 g of CO_2 and 0.1172 g of H_2O were obtained. What is the empirical formula of the compound?

SOLUTION:

Step 1: Begin by calculating the mass of carbon and hydrogen in CO_2 and H_2O.

$$\text{C: } 0.1910 \text{ g CO}_2 \times \frac{12.01 \text{ g C}}{44.01 \text{ g CO}_2} = 0.05212 \text{ g C}$$

$$\text{H: } 0.1172 \text{ g H}_2\text{O} \times \frac{2.02 \text{ g H}}{18.02 \text{ g H}_2\text{O}} = 0.01314 \text{ g H}$$

When these masses of C and H are subtracted from the given total mass of 0.1 g, the remainder must be the mass of O. Thus,

$$\text{mass of O} = (\text{mass of sample}) - (\text{total mass of C and H})$$
$$= 0.100 \text{ g} - (0.05212 + 0.01314) \text{ g} = 0.03474 \text{ g O}$$

Step 2: Convert the masses of C, H, and O to moles of each of the elements.

$$C: 0.05212 \text{ g C} \times \frac{1 \text{ mol C}}{12.01 \text{ g C}} = 0.004340 \text{ mol C}$$

$$H: 0.01311 \text{ g H} \times \frac{1 \text{ mol H}}{1.01 \text{ g H}} = 0.012980 \text{ mol H}$$

$$O: 0.03474 \text{ g O} \times \frac{1 \text{ mol O}}{16.0 \text{ g O}} = 0.002171 \text{ mol O}$$

Step 3: Divide each number of moles by the smallest of the values.

$$C = \frac{0.004340 \text{ mol}}{0.002171 \text{ mol}} = 2.00$$

$$H = \frac{0.012980 \text{ mol}}{0.002171 \text{ mol}} = 5.99$$

$$O = \frac{0.002171 \text{ mol}}{0.002171 \text{ mol}} = 1.00$$

Therefore, the formula is $C_{2.00}H_{5.99}O_{1.00}$, and rounds to **C_2H_6O** which is the empirical formula.

PROBLEM 36
What is the empirical formula of a compound composed of only phosphorous (43.7%) and oxygen?

SOLUTION:
The formula is in the form P_xO_y.
Assuming 100 g of the compound,
$\therefore$ The 43.7% of P = 43.7 g P, and the remaining (100-43.7)% = 56.3 g of O.

$$\text{mol P} = \frac{43.7 \text{ g P}}{31.0 \text{ g / mol P}} = 1.41 \text{ mol P}$$

$$\text{mol O} = \frac{56.3 \text{ g O}}{16.0 \text{ g / mol O}} = 3.52 \text{ mol O}$$

Following the examples in the previous three examples, divide through by the smaller of the two values. This give the values; 1 for P and 2.5 for O. Multiplying these by 2 gives **P_2O_5** as the empirical formula.

PROBLEM 37
A compound is found to contain 2.38 g of carbon, 0.60 g of hydrogen, and 7.03 g of chlorine and no other elements. What is the empirical formula?

SOLUTION:
The formula may be written in the form $C_xH_yCl_z$.

$$\text{mol C} = 2.38 \ \text{g C} \times \frac{1 \ \text{mol}}{12.01 \ \text{g}} = 0.1982 \ \text{mol C} \ = \ x$$

$$\text{mol H} = 0.60 \ \text{g H} \times \frac{1 \ \text{mol}}{1.01 \ \text{g}} = 0.594 \ \text{mol H} \ = \ y$$

$$\text{mol Cl} = 7.03 \ \text{g Cl} \times \frac{1 \ \text{mol}}{35.5 \ \text{g}} = 0.1980 \ \text{mol Cl} \ = \ z$$

Thus $C_xH_yCl_z = C_{0.1982}H_{0.594}Cl_{0.1980}$.
  Dividing through by the smallest of the three values gives $C_{1.002}H_{3.000}Cl_{1.000}$.
  $\therefore$ The empirical formula is **CH_3Cl**.

PROBLEM 38
  Nicotine has an empirical formula of C_5H_7N and a molecular weight of 162.0. What is its molecular formula?

SOLUTION:

$$5C = 5(12.01) = 60.05 \ \text{g}$$
$$7H = 7(1.01) = 7.07 \ \text{g}$$
$$1N = 1(14.0) = 14.0 \ \text{g}$$
$$\text{Total} = 81.12 \ \text{g/mol}$$

$$\text{Number of } C_5H_7N \text{ units per molecule } = \frac{162.0 \ \text{g}}{81.12 \ \text{g}} \approx 2$$

$\therefore$ the molecular formula is $(C_5H_7N_1)_2 = $ **$C_{10}H_{14}N_2$**.

PROBLEM 39
  A colorless compound of nitrogen and oxygen with a molar mass of 92.0 g was found to have an empirical formula of NO_2. What is its molecular formula?

SOLUTION:

$$1N = 1(14.0) = 14.0 \ \text{g}$$
$$2O = 2(16.0) = 32.0 \ \text{g}$$
$$\text{Total} = 46.0 \ \text{g}$$

$$\text{Number of units of } NO_2 \text{ per molecule } = \frac{92.0 \ \text{g}}{46.0 \ \text{g}} = 2$$

$\therefore$ the molecular formula is $(NO_2)_2 = $ **N_2O_4**.

PROBLEM 40
  The hydrocarbon propylene has a molar mass of 42.0 g and contains 14.3% H and 85.7% C. What is its molecular formula?

SOLUTION:
  Step 1: Find the empirical formula:
  Assuming 100 g of sample,

$$C: 85.7 \text{ g C} \times \frac{1 \text{ mol C atoms}}{12.01 \text{ g C}} = 7.14 \text{ mol C atoms}$$

$$H: 14.3 \text{ g H} \times \frac{1 \text{ mol H atoms}}{1.01 \text{ g H}} = 14.16 \text{ mol H atoms}$$

Divide each value by the smaller number of moles.

$$C = \frac{7.14 \text{ mol}}{7.14 \text{ mol}} = 1.0$$

$$H = \frac{14.16 \text{ mol}}{7.14 \text{ mol}} \approx 2.0$$

∴ the empirical formula = **CH$_2$**

Step 2: Determine the molecular formula from the empirical formula and molar mass. Write the molecular formula as $(CH_2)_\eta$, where η is the number of empirical formula units. Molar mass = 42.0 g.

Each CH$_2$ unit has a mass of (12.0 g + 2.0 g) or 14.0 g.

$$\therefore \eta = \frac{42.0 \text{ g}}{14.0 \text{ g}} = 3$$

The molecular formula is **C$_3$H$_6$**.

PROBLEM 41

Hydroquinone, an organic compound commonly used as a photographic developer, has a molecular mass of 110.0 g/mol and a composition of 65.45% C, 5.45% H, and 29.09% O. What is its molecular formula?

SOLUTION:

Step 1: Find the empirical formula:

$$C: 64.45 \text{ g C} \times \frac{1 \text{ mol C atoms}}{12.01 \text{ g C}} = 5.37 \text{ mol C atoms}$$

$$H: 5.45 \text{ g H} \times \frac{1 \text{ mol H atoms}}{1.01 \text{ g H}} = 5.40 \text{ mol H atoms}$$

$$O: 29.09 \text{ g O} \times \frac{1 \text{ mol O atoms}}{16.0 \text{ g O}} = 1.82 \text{ mol O atoms}$$

Divide each value by the smallest number of moles.

$$C = \frac{5.37 \text{ mol}}{1.82 \text{ mol}} \approx 3.0$$

$$H = \frac{5.45 \text{ mol}}{1.82 \text{ mol}} \approx 3.0$$

$$O = \frac{1.82 \text{ mol}}{1.82 \text{ mol}} \approx 1.0$$

∴ the empirical formula = **C_3H_3O**

Step 2: Determine the molecular formula from the empirical formula and molar mass. The molecular formula is in the form $(C_3H_3O)_\eta$. Its molar mass = 110.0 g. The mass of each empirical unit, $C_3H_3O = 55.0$ g.

$$\therefore 55\,\eta = 110.0$$
$$\eta = 110/55 = 2$$

The molecular formula is $(C_3H_3O)_2 = \mathbf{C_6H_6O_2}$.

PROBLEM 42

Aspirin, a well known pain reliever, has a molecular mass of 180 g/mol and a composition of 60.0% C, 4.48% H, and 35.5% O. What is the molecular formula of aspirin?

SOLUTION:

Step 1: Find the empirical formula:

Assume 100 g of the aspirin.

$$C: 60.0 \text{ g C} \times \frac{1 \text{ mol C atoms}}{12.01 \text{ g C}} = 5.00 \text{ mol C atoms}$$

$$H: 4.48 \text{ g H} \times \frac{1 \text{ mol H atoms}}{1.01 \text{ g H}} = 4.44 \text{ mol H atoms}$$

$$O: 35.5 \text{ g O} \times \frac{1 \text{ mol O atoms}}{16.0 \text{ g O}} = 2.22 \text{ mol O atoms}$$

Divide each value by the smallest number of moles.

$$C = \frac{5.00 \text{ mol}}{2.22 \text{ mol}} \approx 2.25$$

$$H = \frac{4.44 \text{ mol}}{2.22 \text{ mol}} \approx 2.0$$

$$O = \frac{2.22 \text{ mol}}{2.22 \text{ mol}} \approx 1.0$$

Step 2: Determine the molecular formula from the empirical formula and the molar mass. The molecular formula is in the form $(C_{2.25}H_2O_1)_\eta$. Its molar mass = 180.0 g.

$$[(12 \times 2.25) + (1.01 \times 2) + (16 \times 1)]\eta = 180.02$$
$$\therefore 45.02\eta = 180.02$$
$$\eta = 180.02 \div 45.02 = 4 \text{ empirical units}$$

∴The molecular formula is $(C_{2.25}H_2O)_4 = \mathbf{C_9H_8O_4}$.

PROBLEM 43

12.00 g of zinc and 6.50 g of sulfur are reacted together to form zinc sulfide. The equation for the reaction is

$$Zn + S \rightarrow ZnS$$

(a) Which is the limiting reactant?
(b) How many grams of ZnS can be formed, based on the amount of the limiting reactant?
(c) How many grams of which reactant will remain unreacted?

SOLUTION PROTOCOL:

In solving this and other "limiting-reactant" problems, follow the steps below:
(a) To identify the limiting reactant, we choose one reactant and calculate the amount of the other that would be needed to give a complete reaction. Then compare the amount needed with the amount available to see if we really have enough.
(b) When we know the limiting reactant, we use it to calculate the amount of the product that can be formed.
(c) If zinc is the limiting reactant, sulfur must be the reactant left over after the reaction is completed. Based on the amount of zinc available we will calculate the amount of sulfur that will react. The amount of sulfur left will be equal to the difference between the amount of sulfur available and the amount used.

SOLUTION:

Convert the number of grams to moles.

$$12.00 \text{ g Zn} \times \frac{1 \text{ mol Zn}}{65.4 \text{ g Zn}} = 0.183 \text{ mol Zn}$$

$$6.50 \text{ g S} \times \frac{1 \text{ mol S}}{32.1 \text{ g S}} = 0.202 \text{ mol S}$$

(a) We can now pick one of the reactants, but it does not matter which one, and find out how many moles of the other are needed for complete reaction. Lets pick zinc for example. We have 0.183 mol Zn, and see from above that Zn and S react in one-to-one mole ratio, we will need 0.183 mol of S to use up all the Zn. Zn is the limiting reactant in this reaction since we have 0.202 mol S, which is in excess. We would have come to the same result had we chosen the sulfur to work with. Based on the one-to-one ratio of our balanced equation above, 0.202 mol S would require 0.202 mol Zn in the reaction mixture, but we only have 0.183 mol S to use up all the Zn. In this case the reaction mixture does not have enough Zn, then Zn must still be the limiting reactant.

(b) Use the amount of Zn (the limiting reactant) to calculate the amount of product formed.

From the coefficients of the equation above we have, 1 mol Zn $\Leftrightarrow$ 1 mol ZnS

We now go from moles Zn to moles ZnS and use the formula mass of ZnS (97.5 g/mol) to convert to grams.

$$1 \text{ mol ZnS} = 97.5 \text{ g ZnS}$$

$$\therefore 0.183 \text{ mol Zn} \times \frac{1 \text{ mol ZnS}}{1 \text{ mol Zn}} \times \frac{97.5 \text{ g ZnS}}{1 \text{ mol Zns}} = 17.80 \text{ g ZnS}$$

(c) Since zinc is the limiting reactant, sulfur will be left over.

$$\text{mol S left} = (0.202 - 0.183) \text{ mol S} = 0.019 \text{ mol S}$$

$$\therefore \text{ the g of S remaining} = 0.019 \text{ mol S} \times \frac{32.1 \text{ g S}}{1 \text{ mol S}} = 0.61 \text{ g S}$$

$\therefore$ after the reaction is finished 0.61 g S is left.

PROBLEM 44

The combustion of ethylene, C_2H_4, in air to form CO_2 and H_2O is given according to the equation below:

$$C_2H_4 + 3O_2 \rightarrow 2CO_2 + 2H_2O$$

How many grams of CO_2 will be formed when a mixture containing 1.93 g C_2H_4 and 5.92 g O_2 is ignited?

SOLUTION PROTOCOL:

Change the amounts of each reactant to moles as before and identify the limiting reactant. Choose one reactant and ask how many moles of the other are needed for complete reaction. Compare the amount needed with the amount actually available and decide which is limiting. Use the number of moles of the limiting reactant actually available to calculate the amount of product formed.

SOLUTION:

First we convert the given amounts of the reactants to moles:

$$1.93 \text{ g } C_2H_4 \times \frac{1 \text{ mol } C_2H_4}{28.0 \text{ g } C_2H_4} = 0.0689 \text{ mol } C_2H_4$$

$$5.92 \text{ g } O_2 \times \frac{1 \text{ mol } O_2}{32.0 \text{ g } O_2} = 0.185 \text{ mol } O_2$$

Let's pick O_2 (we could also have picked C_2H_4 instead) and find out how many moles of the other are needed for complete reaction.

How many moles of C_2H_4 are needed to completely use up O_2? From the equation we have the mole ratio as:

$$1 \text{ mol } C_2H_4 \Leftrightarrow 3 \text{ mol } O_2$$

$$0.185 \text{ mol O}_2 \times \frac{1 \text{ mol C}_2\text{H}_4}{3 \text{ mol O}_2} = 0.0617 \text{ mol C}_2\text{H}_4$$

As seen from the calculation, we need 0.0617 mol C_2H_4, but we have 0.0689 mol C_2H_4, thus we have more C_2H_4, than we need. All the O_2 should be able to react.

$\therefore$ O_2 is the limiting reactant.

The amount of CO_2 formed is calculated using the amount of limiting reactant as:

$$0.185 \text{ mol O}_2 \times \frac{2 \text{ mol CO}_2}{3 \text{ mol O}_2} \times \frac{44.0 \text{ g CO}_2}{1 \text{ mol CO}_2} = 5.43 \text{ g CO}_2$$

$\therefore$ The maximum amount of CO_2 obtained from the reaction is 5.43 g.

PROBLEM 45

Magnesium nitride is prepared by the reaction of magnesium metal with nitrogen gas according to the equation shown below:

$$3 \text{ Mg(s)} + \text{N}_2(\text{g}) \rightarrow \text{Mg}_3\text{N}_2(\text{s})$$

(a) How many grams of the product ($Mg_3N_2(s)$) can be formed when 40.00 g of magnesium reacts with 20.00 g of nitrogen? (b) How many grams of the excess reactant remain after the reaction is complete?

SOLUTION:

(a) As before, we have to identify the limiting reactant. We can do two simple calculations, i.e. the number of moles of $Mg_3N_2(s)$ produced based on two different assumptions.

(1) Assuming Mg is the limiting reactant and N_2 is in excess:

$$\text{mol Mg}_3\text{N}_2 = 40.00 \text{ g Mg} \times \frac{1 \text{ mol Mg}}{24.30 \text{ g Mg}} \times \frac{1 \text{ mol Mg}_3\text{N}_2}{3 \text{ mol Mg}}$$

$$= 0.549 \text{ mol Mg}_3\text{N}_2$$

(2) Assuming N_2 is the limiting reactant and Mg is in excess:

$$\text{mol Mg}_3\text{N}_2 = 20.00 \text{ g N}_2 \times \frac{1 \text{ mol Mg}_3\text{N}_2}{28.0 \text{ g N}_2} = 0.714 \text{ mol Mg}_3\text{N}_2$$

The amount of product in the first calculation (0.549 mol Mg_3N_2) is smaller than that in the second (0.714 mol Mg_3N_2), then magnesium is the limiting reactant. The mass of Mg_3N_2 formed when the reaction is complete is:

$$\text{g Mg}_3\text{N}_2 = 0.549 \text{ mol Mg}_3\text{N}_2 \times \frac{100.93 \text{ g Mg}_3\text{N}_2}{1 \text{ mol Mg}_3\text{N}_2} = 55.41 \text{ g Mg}_3\text{N}_2$$

(b) Having found that the amount of product is 0.549 mol Mg_3N_2, we can calculate how much N_2 in grams was consumed.

$$g\, N_2 \;=\; 0.549 \text{ mol Mg}_3N_2 \times \frac{1 \text{ mol } N_2}{1 \text{ mol Mg}_3N_2} \times \frac{28.0 \text{ g } N_2}{1 \text{ mol } N_2} = 15.37 \text{ g } N_2$$

The mass of N_2 present in excess is:

$$20.00 \text{ g } N_2 \text{ (initial)} - 15.37 \text{ g } N_2 \text{ (consumed)} = 4.6 \text{ g } N_2 \text{ (excess)}$$

PROBLEM 46

How many moles of Fe_3O_4 can be obtained when of 8.40 g Fe is heated with 5.00 g H_2O? Which substance is the limiting reactant? Which substance is in excess?

SOLUTION:

Write a balanced chemical equation, showing the reactants and products.

$$3Fe(s) + 4H_2O(g) \xrightarrow{\;\Delta\;} Fe_3O_4(s) + 4H_2(g)$$

Calculate the moles of Fe_3O_4 that can be formed from each reactant.

$$g \text{ reactant} \rightarrow \text{mol reactant} \rightarrow \text{mol } Fe_3O_4$$

Step 1: Calculate the moles of Fe_3O_4 that can be formed from each reactant.

$$8.4 \text{ g Fe} \times \frac{1 \text{ mol Fe}}{55.8 \text{ g Fe}} \times \frac{1 \text{ mol } Fe_3O_4}{3 \text{ mol Fe}} = 0.050 \text{ mol } Fe_3O_4$$

$$5.0 \text{ g } H_2O \times \frac{1 \text{ mol } H_2O}{18.0 \text{ g } H_2O} \times \frac{1 \text{ mol } Fe_3O_4}{4 \text{ mol } H_2O} = 0.069 \text{ mol } Fe_3O_4$$

Step 2: Determine the limiting reactant. The limiting reactant is Fe because it produces less Fe_3O_4; the H_2O is in excess. The yield of the product is 0.050 mol of Fe_3O_4.

PROBLEM 47

How many grams of silver iodide, AgI, can be formed when solutions containing 50.0 g of MgI_2 and 100.0 g $AgNO_3$ are reacted together? How many grams of the excess reactant remain unreacted?

SOLUTION:

Write a balanced chemical equation, showing the reactants and products.

$$MgI_2(aq) + 2\, AgNO_3(aq) \rightarrow 2\, AgI\!\downarrow + Mg(NO_3)_2(aq)$$

Step 1: Calculate the grams of AgI that can be formed from each reactant.

$$g \text{ reactant} \rightarrow \text{mol reactant} \rightarrow \text{mol AgI} \rightarrow g \text{ AgI}$$

$$50.0 \text{ g } MgI_2 \times \frac{1 \text{ mol } MgI_2}{278.1 \text{ g } MgI_2} \times \frac{2 \text{ mol AgI}}{1 \text{ mol } MgI_2} \times \frac{234.7 \text{ g AgI}}{1 \text{ mol AgI}} = 84.39 \text{ g AgI}$$

$$100.0 \text{ g AgNo}_3 \times \frac{1 \text{ mol AgNO}_3}{169.8 \text{ g AgNO}_3} \times \frac{2 \text{ mol AgI}}{2 \text{ mol AgNO}_3} \times \frac{234.7 \text{ g AgI}}{1 \text{ mol AgI}} = 138.2 \text{ g AgI}$$

Step 2: Determine the limiting reactant. The limiting reactant is MgI_2 because it gives less AgI. $AgNO_3$ is in excess. The yield is 84.39 g AgI.

Step 3: Calculate the grams of unreacted $AgNO_3$, i.e., calculate the grams of $AgNO_3$ that will react with 50.0 g of MgI_2.

$$\text{g MgI}_2 \rightarrow \text{mol MgI}_2 \rightarrow \text{mol AgNO}_3 \rightarrow \text{g AgNO}_3$$

$$50.0 \text{ g MgI}_2 \times \frac{1 \text{ mol MgI}_2}{278.1 \text{ g MgI}_2} \times \frac{2 \text{ mol AgNO}_3}{1 \text{ mol MgI}_2} \times \frac{169.8 \text{ g AgNO}_3}{1 \text{ mol AgNO}_3} = 61.06 \text{ g AgNO}_3$$

In conclusion, 61.06 g of $AgNO_3$ reacted with 50.0 g of MgI_2. The amount of $AgNO_3$ that remains unreacted is (100.0-61.06) g $AgNO_3$ = 38.94 g$AgNO_3$ unreacted. The final mixture will contain 84.39 g AgI(s), 38.94 g $AgNO_3$ and $Mg(NO_3)_2$ in solution.

PROBLEM 48

Carbon tetrachloride was prepared by reacting 95.50 g of carbon disulfide and 95.50 g of chlorine. Calculate the percent yield if 60.50 g of CCl_4 was obtained from the reaction.

$$CS_2 + 3Cl_2 \rightarrow CCl_4 + S_2Cl_2$$

SOLUTION:

To calculate the theoretical yield, we need to determine the limiting reactant. Then we can compare this amount with the 60.5 g of CCl_4 actual yield to calculate the percent yield.

Step 1: Determine the theoretical yield. Calculate the grams of CCl_4 that can be formed from each reactant.

$$\text{g reactant} \rightarrow \text{mol reactant} \rightarrow \text{mol CCl}_4 \rightarrow \text{g CCl}_4$$

$$95.5 \text{ g CS}_2 \times \frac{1 \text{ mol CS}_2}{76.10 \text{ g CS}_2} \times \frac{1 \text{ mol CCl}_4}{1 \text{ mol CS}_2} \times \frac{153.80 \text{ g CCl}_4}{1 \text{ mol CCl}_4} = 193.00 \text{ g CCl}_4$$

$$95.5 \text{ g Cl}_2 \times \frac{1 \text{ mol Cl}_2}{70.90 \text{ g Cl}_2} \times \frac{1 \text{ mol CCl}_4}{3 \text{ mol Cl}_2} \times \frac{153.80 \text{ g CCl}_4}{1 \text{ mol CCl}_4} = 69.00 \text{ g CCl}_4$$

The limiting reactant is Cl_2 because it gives less CCl_4. The CS_2 is in excess. The theoretical yield is 69.0 gCCl_4.

Step 2: Calculate the percent yield. According to the equation, 69.00 g of CCl_4 is the maximum amount or theoretical yield of CCl_4 possible from 95.50 g of Cl_2 and the actual yield is 60.50 g of CCl_4.

$$\%\text{Yield} = \frac{60.50 \text{ g CCl}_4}{69.00 \text{ g CCl}_4} = 87.7\%$$

PROBLEM 49

Silver bromide was prepared by reacting 195.00 g of magnesium bromide and an adequate amount of silver nitrate. Calculate the percent yield if 370.00 g of silver bromide was obtained from the reaction.

$$MgBr_2 + 2AgNO_3 \rightarrow Mg(NO_3)_2 + 2AgBr$$

SOLUTION:

Step 1: Determine the theoretical yield. Calculate the grams of AgBr that can be formed.

$$195.0 \text{ g MgBr}_2 \times \frac{1 \text{ mol MgBr}_2}{184.1 \text{ g MgBr}_2} \times \frac{2 \text{ mol AgBr}}{1 \text{ mol MgBr}_2} \times \frac{187.7 \text{ g AgBr}}{1 \text{ mol AgBr}} = 397.60 \text{ g AgBr}$$

The theoretical yield is 397.60 g AgBr.

Step 2: Calculate the percent yield. According to the equation, 397.6 g AgBr is the maximum amount of AgBr possible from 195.00 g MgBr$_2$. Actual yield is 370.00 g AgBr.

$$\%\text{Yield} = \frac{370.0 \text{ g AgBr}}{397.6 \text{ g AgBr}} = 93.0\%$$

PROBLEM 50

Aluminum oxide was prepared by heating 220.00 g of chromium(II)oxide with 120.00 g of aluminum. Calculate the percent yield if 100.00 g of aluminum oxide was obtained.

$$2Al + 3CrO \rightarrow Al_2O_3 + 3Cr$$

SOLUTION:

Step 1: Determine the theoretical yield. Calculate the grams of Al$_2$O$_3$ that can be formed from each reactant.

$$\text{g reactant} \rightarrow \text{mol reactant} \rightarrow \text{mol Al}_2O_3 \rightarrow \text{g Al}_2O_3$$

$$220.0 \text{ g CrO} \times \frac{1 \text{ mol CrO}}{68.0 \text{ g CrO}} \times \frac{1 \text{ mol Al}_2O_3}{3 \text{ mol CrO}} \times \frac{102.0 \text{ g Al}_2O_3}{1 \text{ mol Al}_2O_3} = 110.0 \text{ g Al}_2O_3$$

$$120.0 \text{ g Al} \times \frac{1 \text{ mol Al}}{27.0 \text{ g Al}} \times \frac{1 \text{ mol Al}_2O_3}{2 \text{ mol Al}} \times \frac{102.0 \text{ g Al}_2O_3}{1 \text{ mol Al}_2O_3} = 226.7 \text{ g Al}_2O_3$$

The limiting reactant is CrO because it gives less Al$_2$O$_3$. The Al is in excess. The theoretical yield is 110.0 g Al$_2$O$_3$.

Step 2: Calculate the percent yield.

According to the equation, 110.0 g of Al$_2$O$_3$ is the maximum amount or theoretical yield of Al$_2$O$_3$ possible from 220.0 g of CrO and the actual yield is 100.0 g of Al$_2$O$_3$.

$$\%\text{Yield} = \frac{100.0 \text{ g } Al_2O_3}{110.0 \text{ g } Al_2O_3} \times 100\% = 90.9\%$$

PROBLEM 51

Freon-12, CCl_2F_2, the refrigerant gas, is made as shown in the reaction below:

$$3CCl_4 + 2SbF_3 \rightarrow 3CCl_2F_2 + 2SbCl_3$$

In a certain experiment 14.60 g of SbF_3 was allowed to react with an excess of CCl_4. After the reaction was completed, 8.62 g of freon-12 was isolated from the mixture.
(a) What was the theoretical yield of CCl_2F_2 in grams from this reaction?
(b) What was the actual yield of CCl_2F_2 in grams?
(c) What was the percentage yield of CCl_2F_2 in this experiment?

SOLUTION:

Step 1: Determine the theoretical yield. Calculate the grams of CCl_2F_2 that can be formed.

$$14.60 \text{ g } SbF_3 \times \frac{1 \text{ mol } SbF_3}{178.7 \text{ g } SbF_3} \times \frac{3 \text{ mol } CCl_2F_2}{2 \text{ mol } SbF_3} \times \frac{120.9 \text{ g } CCl_2F_2}{1 \text{ } CCl_2F_2} = 14.82 \text{ g } CCl_2F_3$$

The theoretical yield is 14.82 g CCl_2F_2
Step 2: Calculate the percent yield. According to the equation, 14.82 g CCl_2F_2 is the maximum amount of CCl_2F_2 possible from 14.60 g SbF_3. Actual yield is 8.62 g CCl_2F_2.

$$\%\text{Yield} = \frac{8.62 \text{ g } CCl_2F_2}{14.82 \text{ g } CCl_2F_2} \times 100\% = 58.2\%$$

∴ The answers are:
(a) 14.8 g CCl_2F_2
(b) 8.62 g CCl_2F_2
(c) 58.2%

PROBLEM 52

Given the reaction:

$$2SO_2 + 1O_2 \rightarrow 2SO_3$$

and the rate of disappearance of O_2 as 2.5×10^{-2} M/sec, calculate the rate of disappearance SO_2 and the rate of appearance of SO_3.

SOLUTION:

First we write a mathematical expression for the rate of disappearance of the reactants and the appearance of product:

Let the rate of disappearance of SO_2 be $-\dfrac{\Delta(SO_2)}{\Delta t}$

Similarly, let the rate of appearance of SO_3 be $+\dfrac{\Delta(SO_3)}{\Delta t}$

Then, $\dfrac{-\dfrac{\Delta(SO_2)}{\Delta t}}{2} = \dfrac{-\dfrac{\Delta(O_2)}{\Delta t}}{1} = 2.5 \times 10^{-2}$ M/sec

$-\dfrac{\Delta(SO_2)}{\Delta t} = 2 \times (2.5 \times 10^{-2} \text{ M / sec}) = 5.0 \times 10^{-2} \text{ M / sec}$

Also, $\dfrac{-\dfrac{\Delta(O_2)}{\Delta t}}{1} = \dfrac{+\dfrac{\Delta(SO_3)}{\Delta t}}{2}$

Then, $+\dfrac{\Delta(SO_3)}{\Delta t} = 2 \times (2.5 \times 10^{-2} \text{ M / sec}) = 5.0 \times 10^{-2} \text{ M / sec.}$

$\therefore$ the rate in both cases $= 5.0 \times 10^{-2}$ M/sec

PROBLEM 53
Use the data given below for the reaction of HCl gas with O_2 gas to determine the rate law and the rate constant.

Experiment	Initial concentration (mol/L)		Initial rate of formation of product (mol L^{-1} s^{-1})
	[HCl]	[O$_2$]	
1	1.50	9.80	5.4×10^{-3}
2	3.00	9.80	10.8×10^{-3}
3	3.00	19.6	2.2×10^{-2}

(a) What is the order of the reaction with respect to each of the reactants?
(b) What is the overall order of the reaction?
(c) What is the value of the rate constant?
(d) What is the overall rate law for the reaction?

SOLUTION:
(a) We use experiments 1 and 2 where [O$_2$] is constant to obtain the order with respect to [HCl]. In going from 1 to 2, the rate is seen to double as [HCl] doubles. So order in [HCl] = 1.

We then use experiments 2 and 3 where [HCl] is constant to obtain the order with respect to [O$_2$]. In going from 2 to 3, the rate is seen to double also as [O$_2$] doubles. The order in [O$_2$] is also = 1.

(b) The overall order of the reaction is 1+1 = 2.
(c) The rate constant is calculated using:

$$\text{rate} = k[HCl]^1[O_2]^1$$

$$k = \frac{\text{rate}}{[HCl]^1[O_2]^1} = \frac{5.4 \times 10^{-3} \text{ mol L}^{-1}\text{s}^{-1}}{(1.50)(9.80)(\text{mol}^2\text{L}^{-2})} = 3.7 \times 10^{-4} \text{ mol}^{-1} \text{ Ls}^{-1}$$

(d) Rate= 3.7×10^{-4} mol^{-1} L s^{-1}[HCl]^1[O$_2$]^1

PROBLEM 54

Use the data given below for the reaction of bromine gas with nitric oxide gas to determine the rate law and the rate constant.

$$2NO(g) + Br_2(g) \rightarrow 2NOBr(g)$$

Experiment	Initial concentration (mol/L)		Initial rate of formation of NOBr (mol L^{-1} s^{-1})
	NO	Br$_2$	
1	0.010	0.010	1.2×10^{-1}
2	0.010	0.020	2.4×10^{-1}
3	0.010	0.030	3.6×10^{-1}
4	0.020	0.010	4.8×10^{-1}
5	0.030	0.010	10.8×10^{-1}

SOLUTION:

The rate law for the reaction has the form

$$\text{Rate} = k[NO]^n[Br_2]^m$$

To determine either n or m, we will study how the rate changes when the concentration of one reactant varies while that of the other reactant stays the same. Hence, when the NO concentration is held constant, we can figure out how changes in the Br$_2$ concentration affect the rate and thus determine what the value of n must be. The value of m is determined in a similar way.

With this in mind, in experiments 1 to 3, the concentration of NO is constant and that of Br$_2$ is varied. When the concentration of Br$_2$ is doubled (experiments 1 and 2), the rate is increased by a factor of 2, when it is tripled (experiments 1 and 3), the rate is increased by a factor of 3. Since the concentration of Br$_2$ is directly proportional to rate, it must be to the first power in the rate law. Therefore, m = 1

Comparing experiments 1 and 4, we see that when Br$_2$ concentration is held constant, the rate increases by a factor of four when NO concentration is multiplied by 2. Also, raising the concentration of NO by a factor of 3 causes a ninefold increase in the rate (experiments 1 and 5). This means that the exponent on the NO concentration in the rate law must be 2. Therefore n = 2.

$$\therefore \text{Rate} = k[NO]^2[Br_2]^1$$

Overall order of the reaction = 2 +1 = 3.

The rate constant can be calculated from any data from the experiments. Using the data from experiment 4, we have

$$4.8 \times 10^{-1} \text{ mol L}^{-1}\text{s}^{-1} = k(0.020 \text{ mol} \cdot \text{L}^{-1})^2 (0.010 \text{ mol} \cdot \text{L}^{-1}) = k(4.0 \times 10^{-6} \text{ mol}^3 \cdot \text{L}^{-3})$$

$$k = \frac{4.8 \times 10^{-1} \text{ mol} \cdot \text{L}^{-1} \cdot \text{s}^{-1}}{4.0 \times 10^{-6} \text{ mol}^3 \cdot \text{L}^{-3}} = 1.2 \times 10^5 \text{ mol}^2 \cdot \text{L}^2 \cdot \text{s}^{-1}$$

Chapter 3

Properties of Gases and Colligative Properties of Solutions

3.1. Properties of Gases

A. Principles

There are three different physical forms or states of matter; solid, liquid, and gas. In this section, we will deal only with the gaseous state. In gaseous state the intermolecular forces of attraction (the attraction that one molecule experiences toward others around it) are weak. These weak forces allow for rapid and independent movement of the gaseous molecules and cause the physical behavior of a gas to be nearly independent of its chemical composition. Gas molecules move with very high velocities and have high kinetic energy. Because of the high velocities of their molecules, mixtures of gases are uniformly distributed within the container in which they are confined. The behavior of a gas is controlled by its volume, pressure, temperature, and number of moles. These variables are very important and we will now take a close look at them.

B. Volume and Pressure

When introduced into a vessel, the molecules of a gas move freely and occupy the entire volume of the vessel. As a result, the volume of a gas is given simply by specifying the volume of the vessel in which it is held. The pressure of a gas, defined as force per unit area, is produced by gas molecules colliding with the walls of the vessel containing it. The number of collisions can be increased by increasing the number of molecules present. For an ideal gas, if we double the number of molecules, the frequency of collisions and the pressure also double. When the temperature and volume are kept constant at 0°C (273 K) and 22.4 L, the pressure is directly proportional to the number of moles or molecules of the gas present. Let us look at the quantitative aspects of gas behavior.

Example: See problem 55.

C. Boyle's Law

We already know that the behavior of a gas is controlled by its volume, pressure, temperature and number of moles. In 1662 Robert Boyle, an Irish chemist and physicist, reported that if the temperature (T) of a gas sample is held constant, its volume (V) is inversely proportional to the pressure (P) exerted on it. This relationship of P and V is known as **Boyle's law**. The law can be expressed mathematically as:

$$V \propto \frac{1}{P} \tag{55}$$

This means that when the pressure on a gas is increased, its volume will decrease, and vice versa. The proportionality in equation (55) can be changed to an equality by inserting a proportionality constant,

$$V = \frac{Constant}{P} \qquad \text{Thus, PV = Constant} \tag{56}$$

Since the product of pressure and volume is the same number all the time, we can write that

$$P_1 V_1 = P_2 V_2 \tag{57}$$

Examples: See problems 56–57.

D. Charles' Law

Around 1787, hot-air ballooning was popular in France. It was about this time that a French chemist and balloon enthusiast Jacques A. Charles observed that several gases expand by the same fractional amount when heated through the same temperature interval. It was also found that when a given volume of any gas initially at $0°C$ was cooled by $1°C$, the volume decreased by $1/273.15$; if cooled by $2°C$, the volume decreased by $2/273.15$; if cooled by $10°C$, the volume decreased by $10/273.15$; and so on. Because each degree of cooling reduced the volume by $1/273.15$, it was theorized that any quantity of any gas would have zero volume if cooled to $-273.15°C$. Of course no real gas can be cooled to $-273.15°C$ because it will liquefy before that temperature is reached. However, $-273°C$ (for most purposes we need only three significant figures) is referred to as **absolute zero temperature**. It corresponds to the zero point on the Kelvin temperature scale. This is also called the **absolute zero** scale.

Charles' law states that at constant pressure, the volume of a fixed mass of any gas is directly proportional to the absolute temperature. It is expressed mathematically as:

$$V \propto T \quad \text{(at constant pressure)}$$

This can be written as:

$$V = kT \tag{58}$$

$$\frac{V}{T} = k \quad \text{(at constant pressure)} \tag{59}$$

where k is a constant for a fixed mass of gas. Note that T stands for absolute temperature and will always be used whenever the numerical value of temperature is used in calculations involving the volume and pressure of a gas. If the temperature in any problem is given initially in $°C$ ($T_{°C}$), it must be converted to degree Kelvin (T_K) using equation 60 below:

$$T_k = T_{°C} + 273 \tag{60}$$

We can use this law to relate the volume of a gas at two different temperatures as follows:

$$\frac{V_1}{T_1} = k = \frac{V_2}{T_2} \tag{61}$$

$$\therefore \ \frac{V_1}{T_1} = \frac{V_2}{T_2} \tag{62}$$

where V_1 and T_1 are one set of conditions and V_2 and T_2 are another set of conditions.

Examples: See problems 58–59.

E. Gay-Lussac's Law

In section 3.1A, we stated that the variables needed to define the state of a gas were volume, pressure, temperature, and moles. Joseph Gay-Lussac, a contemporary of Charles, independently studied how the pressure of a gas depended on its temperature when the volume was held constant. From his experiments he found a proportionality involving pressure and absolute temperature at constant volume and stated as follows: "**The pressure of a fixed mass of gas held at constant volume, is directly proportional to the absolute temperature.**" This relationship is called the **Gay-Lussac's law**. Quantitatively, it is written:

$$P \propto T \quad \text{(at constant volume), or}$$

$$\frac{P}{T} = \text{constant} \tag{63}$$

And thus,

$$\frac{P_1}{T_1} = k = \frac{P_2}{T_2} \tag{64}$$

$$\therefore \ \frac{P_1}{T_1} = \frac{P_2}{T_2} \tag{65}$$

where P_1 and T_1 are one set of conditions and P_2 and T_2 are another set of conditions.

Examples: See problems 60–61.

F. The Combined Gas Law

The three gas laws above can be combined into a single law known as the **combined gas law**, and is expressed as:

$$\frac{PV}{T} = \text{constant} \tag{66}$$

For problem solving this equation is usually written in the form:

$$\frac{P_1 V_1}{T_1} = \frac{P_2 V_2}{T_2}$$

(67)

where P_1, V_1, and T_1 are one set of conditions, and P_2, V_2, and T_2 are a another set of conditions. This equation can be used to solve for any of the six variables.

Examples: See Problems 62–66.

G. Avogadro's Law

Around 1811, an Italian chemist Amedeo Avogadro made a simple but significant generalization that, equal volumes of gases at the same temperature and pressure contain the same number of "Particles." This observation is called **Avogadro's law**. Stated mathematically, this gives

$$\frac{V}{n} = \text{constant}$$

(68)

where V is the volume of the gas and n is the number of moles of the gas particles. An alternative representation of this law is:

$$\frac{V_1}{n_1} = \text{constant} = \frac{V_2}{n_2}$$

(69)

$$\therefore \frac{V_1}{n_1} = \frac{V_2}{n_2}$$

(70)

where V_1 and n_1 are one set of conditions and V_2 and n_2 are another set of conditions.

Example: See problem 67.

H. The Ideal Gas

The combined gas law indicates that for a gas sample, the ratio PV/T is a constant. If we plot the values of the ratio of PV/T as a function of pressure, we obtain a **horizontal line** as shown in Figure 3.1. A hypothetical gas that would obey the combined gas law over all ranges of pressure and temperature and gives the **horizontal line** in the graph, is called an **ideal gas**. However a real gas, as opposed to an ideal gas, deviates from the ideal behavior and does not yield a horizontal line when PV/T is plotted against P (Figure 3.1). Nevertheless, at relatively low pressures (including atmospheric pressure) and at relatively high temperatures, many gases come quite close to being ideal. The combined gas law works satisfactorily for most calculations unless otherwise stated.

I. The Ideal Gas Law

We have seen that for a fixed mass of gas (assuming ideal behavior)

$$\frac{PV}{T} = \text{constant}$$

(71)

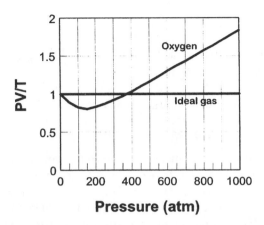

Figure 3.1. **A graph of PV/T versus pressure for an ideal gas and for one mole of oxygen at 0°C.**

The constant in this equation is only a constant if we keep the amount of the gas constant. The value of the constant is directly proportional to the number of moles of the gas. The above equation then becomes

$$\frac{PV}{T} = nx \qquad (72)$$

where x is a combination of two proportionality constants. The x is known as the **universal gas constant,** and is represented by the symbol R. When the pressure is expressed in atmospheres and the volume in liters, R has the value of $0.08206 \, \text{L·atm·K}^{-1}\text{·mol}^{-1}$. By substituting R into equation 72, we obtain:

$$\frac{PV}{T} = nR \qquad (73)$$

Rearranging equation 73 yields,

$$PV = nRT \qquad (74)$$

Equation 74 is known as the **ideal gas law.**

Examples: See problems 68–74.

J.   Standard Temperature, Pressure, and Molar Volume

To compare volumes of gases, a universal reference point of temperature and pressure was selected and called "standard condition" or "standard temperature and pressure" (abbreviated **STP**). Standard temperature is 273 K (0°C), and standard pressure is 1 atm or 760 torr or 760 mm Hg. Suppose that in comparing different gases at STP, we use Avogadro's number as the number of molecules present, then Avogadro's hypothesis states that under these conditions, one mole of any gas occupies the same volume. From the ideal gas law, this volume is given by:

$$V = \frac{nRT}{P} \tag{75}$$

$$= \frac{(1.00 \text{ mol})(0.08206 \text{ L} \cdot \text{atm} \cdot \text{K}^{-1} \cdot \text{mol}^{-1})(273 \text{ K})}{1.000 \text{ atm}} = 22.4 \text{ L} \tag{76}$$

The volume 22.4 L is the **molar volume of any gas** at STP.

K.   Dalton's Law of Partial Pressures

When gases that do not react chemically are mixed in a container, the pressure exerted by each gas is the same as it would be if it were the only gas in the container. The pressure exerted by each gas is called the partial pressure of that gas. As observed by John Dalton, the total or combined pressure exerted by all the gases present, is equal to the sum of the partial pressures of each gas. This relation is known as **Dalton's law of partial pressures** and is expressed as:

$$P_{TOTAL} = P_1 + P_2 + P_3 + \ldots \tag{77}$$

where the subscripts 1, 2, 3, etc., represent the individual gases, and P_1, P_2, P_3, etc, represents their corresponding partial pressures. Assuming each gas behaves ideally, its partial pressure can be calculated from the ideal gas law:

$$P_1 = \frac{n_1 RT}{V}, P_2 = \frac{n_2 RT}{V}, P_3 = \frac{n_3 RT}{V} \tag{78}$$

The total pressure of the mixture, P_{TOTAL}, can be represented as

$$P_{TOTAL} = P_1 + P_2 + P_3 + \ldots = \frac{n_1 RT}{V} + \frac{n_2 RT}{V} + \frac{n_3 RT}{V} \ldots$$

$$= (n_1 + n_2 + n_3 \ldots \ldots)\left(\frac{RT}{V}\right) = n_{(total)}\left(\frac{RT}{V}\right) \tag{79}$$

Examples: See problems 74–77.

L.   Mole Fraction (χ)

Mole fraction (χ) is the ratio of the number of moles of a given component in a mixture to the total number of moles of all components in the mixture. The partial pressure of a gas in a mixture of gases is related to the total pressure by its mole fraction. For example, in a mixture of gas A and gas B,

$$\text{Mole fraction A} = \frac{\text{number of moles A}}{\text{number of moles A} + \text{number of moles B}}$$

Substituting the respective symbols,

$$\chi_A = \frac{n_A}{n_A + n_B} \tag{80}$$

If P_A is the partial pressure of A and P_T is the total pressure, then

$$P_A = \chi_A P_T \tag{81}$$

This is **Raoult's law** (see also equation 85).

Example: See problem 78.

3.2. Colligative Properties of Solutions

A. Principles

In chapters 1 and 2, we addressed some aspects of solution chemistry. We now turn our attention to what is collectively called colligative properties of solutions. The **colligative properties** are properties that depend only upon the number of solute particles in a solution and not on the nature of those particles. Because of their direct relationship to the number of solute particles, the colligative properties are very useful for characterizing the nature of solute after it is dissolved in a solvent and for determining molar masses of substances. There are four colligate properties; **vapor pressure lowering**, **osmotic pressure**, **freezing point depression** and **boiling point elevation**. The discussions on freezing point depression and boiling point elevation will be restricted to un-ionized and non-volatile substances.

Both the freezing point depression and the boiling point elevation are directly proportional to the number of moles of solute per kilogram of solvent. The concentration term, molality (section 1.2B), is used when dealing with these two colligative properties. For un-ionized substances, the colligative properties of their solution are directly proportional to their molality.

B. Boiling Point Elevation

The normal boiling point of a liquid is the temperature at which the vapor pressure of the liquid equals the external pressure exerted on the liquid. When a nonvolatile solute is dissolved in a solvent, the vapor pressure of the solvent is lowered. Such a solution, therefore must be heated to higher temperature than the pure solvent before it equalizes the external pressure. Thus the boiling point is elevated relative to the pure solvent. The magnitude of the boiling point elevation depends on the concentration of the solute. The elevation in boiling point can be represented by the equation

$$\Delta T_b = K_b m_{solute} \tag{82}$$

where ΔT_b is the boiling point elevation, or the difference between the boiling point of the solution and that of the pure solvent, K_b is a constant that is characteristic of the solvent and is called the modal boiling point elevation constant, and m_{solute} is the molality of the solute in the solution.

Examples: See problems 79–81.

C. Freezing Point Depression

Freezing point is the temperature at which a liquid freezes; i.e., the temperature at which the liquid comes into equilibrium with solid. For a pure substance the freezing point and melting point are the same. When a solute is dissolved in a solvent, the freezing point of the solution is lower than that of the pure solvent. The equation for freezing point depression is analogous to that for the boiling point elevation and is written as:

$$\Delta T_f = K_f m_{solute} \tag{83}$$

where ΔT_f is the freezing point depression, or the difference between the freezing point of the solvent and that of the solution, K_f is a constant that is characteristic of a particular solvent and is called the modal freezing point depression constant. The freezing point depression, like the boiling point elevation is used to determine molar masses and to characterize solutions.

Examples: See problems 82–83.

D. Osmotic Pressure; Principles and Applications

There are a few things to note when dealing with osmosis. Firstly, when a solution is allowed to be in contact with a membrane and the size of the solute molecules is larger than the diameter of the pores in the membrane, the solute molecules will stay behind but the solvent molecules will pass through the membrane. However, if the size of the solute molecules is smaller than the pores in the membrane, the solute molecules will also flow across the membrane. **A membrane that allows only the passage of solvent molecules and not the solute, is called a semipermeable membrane, and the flow of solvent across a semipermeable membrane is called osmosis.**

Secondly, if a semipermeable membrane separates two identical solutions, solvent molecules move in both directions at the same rate and there is no net osmosis. Thus, the two sides of the membrane are at dynamic equilibrium. However the situation is different when the concentration of the solution on one side of the membrane is different from the concentration of the solution on the other side. Consider a setup in which a balloon made from a semipermeable membrane filled with a 0.8 M aqueous glucose solution is immersed in a 0.1 M aqueous glucose solution in a beaker (Figure 3.2). The water molecules flow from

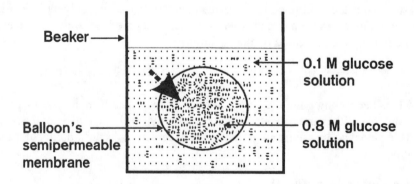

Beaker →

0.1 M glucose solution

Balloon's semipermeable membrane

0.8 M glucose solution

Figure 3.2. Illustration of the principles of osmotic pressure. A semipermeable membrane balloon filled with a 0.8 M glucose solution is immersed in a 0.1 M glucose solution. The resultant concentration difference generated an osmotic pressure that drives the flow of solvent from the 0.1 M solution into the 0.8 M one. The large arrow indicates the direction of solvent flow.

the more dilute 0.1 M solution to the more concentrated 0.8 M solution as the system goes to equilibrium. The glucose molecules do not flow at all because their size is larger than the pore diameter of the membrane. The glucose concentration of the solution inside the balloon will decrease as water flows in, and the glucose concentration of the solution in the beaker will increase as water flows out of the beaker and into the balloon. Likewise, the volume of the solution inside the balloon will increase as water flows in, and the volume of the solution in the beaker will decrease as water flows out of it and into the balloon.

What can be done to increase the rate of solvent flow across the membrane? If we increase the pressure on the side in contact with the 0.8 M solution, this will increase the rate of solvent flow. **The pressure increase needed to equalize the transfer rate is called osmotic pressure (Π).** Osmotic pressure, therefore, is the difference in pressure between the two sides. Pressure is exerted on both sides of the membrane, and Π is the extra pressure that must be exerted on the side in contact with the less concentrated solution, in order to maintain a dynamic equilibrium. **In other words, Π is the pressure required to stop osmosis from occurring between the solution and pure water.**

A small concentration of solute produces a relatively large osmotic pressure. This makes osmotic pressure a sensitive colligative property for use in characterizing solutions and determining their molar masses. Experimental evidence shows that osmotic pressure is proportional to both concentration (expressed in molarity, M) and temperature (expressed in Kelvin, T). Thus,

$$\prod = MRT \qquad\qquad (84)$$

Examples: See problems 84–86.

E. Vapor Pressure

The vapor pressure of a solution is influenced by the presence of a solute. If a nonvolatile solute is dissolved in a liquid solvent, the solvent's vapor pressure is lowered. The vapor pressure of the solvent above a solution ($P_{solution}$) is the product of the vapor pressure of the pure solvent ($P^\circ_{solvent}$) and the mole fraction of the solvent in the solution ($\chi_{solvent}$). This relationship is called **Raoult's law** and in equation form is

$$P_{solution} = \chi_{solvent} P^\circ_{solvent} \qquad\qquad (85)$$

The presence of a solute in a solvent means that $\chi_{solvent}$ is less than one, and $P_{solution}$ is less than $P^\circ_{solvent}$. Thus the solute has lowered the vapor pressure of the solvent. Raoult's law is strictly followed only in an ideal solution, but often works well for non-ideal solutions that are sufficiently dilute.

Example: See problem 87.

3.3. Practical Examples

PROBLEM 55

A sample of hydrogen gas (H_2) has a pressure of 3.0 atm and a volume of 22.4 L at 0°C. (a) Calculate the moles of H_2 molecules present. (b) How many moles are present when the pressure is doubled to 6.0 atm?

SOLUTION:

(a) Using the ideal gas law (equation 74); $PV = nRT$, and solving for n gives

$$n = \frac{PV}{RT}$$

In this case $P = 3.0$ atm, $V = 22.4$ L, $T = 0°C + 273$ K, and $R = 0.08206$ L·atm·K^{-1}·mol^{-1}. Thus

$$n = \frac{(3.0 \text{ atm})(22.4 \text{ L})}{(0.08206 \text{ L} \cdot \text{atm} \cdot \text{K}^{-1} \cdot \text{mol}^{-1})(273 \text{ K})} = 3.0 \text{ mol}$$

(b) When the pressure is doubled to 6 atm, the value of $n = (2 \times 3)$ mol = 6 mol.

PROBLEM 56

What volume will 3.50 L of a gas occupy if the pressure is changed from 760 mm Hg to 635 mm Hg?

SOLUTION:

Use Boyle's law in the form of equation 57.

Step 1: Organize the given information.

$P_1 = 760$ mm Hg $\qquad\qquad\qquad\qquad$ $V_1 = 350$ L

$P_2 = 635$ mm Hg $\qquad\qquad\qquad\qquad$ $V_2 = ?$

Step 2: Write and solve equation for the unknown.

$$P_1V_1 = P_2V_2$$

$$V_2 = V_1 \times \frac{P_1}{P_2}$$

Step 3: Substitute the given information into the equation and solve.

$$V_2 = 350 \text{ L} \times \frac{760 \text{ mm Hg}}{635 \text{ mm Hg}} = 4.19 \text{ L}$$

PROBLEM 57

A gas occupies a volume 400.0 mL at 500.5 torr pressure. To what pressure must the gas be subjected to in order to change the volume to 150.0 mL?

SOLUTION:

Use Boyle's law in the form of equation 57.

Step 1: Organize the given information.

$P_1 = 500.50$ torr $\qquad\qquad\qquad\qquad$ $V_1 = 400.0$ mL

$P_2 = ?$ $\qquad\qquad\qquad\qquad\qquad\qquad$ $V_2 = 150.0$ mL

Step 2: Write and solve equation for the unknown.

$$P_1V_1 = P_2V_2, \qquad \therefore P_2 = P_1 \times \frac{V_1}{V_2}$$

Step 3: Substitute the given information into the equation and solve.

$$P_2 = 500.5 \text{ torr} \times \frac{400.0 \text{ mL}}{150.0 \text{ mL}} = 1335 \text{ torr}$$

PROBLEM 58

3.5 liters of hydrogen at -20°C are allowed to warm to a temperature of 30°C. What is the volume at 30°C if the pressure remains constant?

SOLUTION:

Use Charles' Law in the form of equation 62.

Step 1: Organize the given information.

$$V_1 = 3.50 \text{ L} \qquad\qquad T_1 = -20°C = 253 \text{ K}$$

$$V_2 = ? \qquad\qquad T_2 = 30°C = 303 \text{ K}$$

Step 2: Write and solve equation for the unknown.

$$\frac{V_1}{T_1} = \frac{V_2}{T_2}, \quad \therefore V_2 = V_1 \times \frac{T_2}{T_1}$$

Step 3: Substitute the given information into the equation and solve.

$$V_2 = 3.50 \text{ L} \times \frac{303 \text{ K}}{253 \text{ K}} = 4.19 \text{ L}$$

PROBLEM 59

A sample of gas at 45°C and 2.5 atm has a volume of 2.58 L. What volume will this gas occupy at 68°C and 2.5 atm?

SOLUTION:

Use Charles' Law in the form of equation 62.

Step 1: Organize the given information.

$$V_1 = 2.58 \text{ L} \qquad\qquad T_1 = 45°C + 273 = 318 \text{ K}$$

$$V_2 = ? \qquad\qquad T_2 = 68°C + 273 = 341 \text{ K}$$

Step 2: Write and solve equation for the unknown.

$$\frac{V_1}{T_1} = \frac{V_2}{T_2}, \quad \therefore V_2 = V_1 \times \frac{T_2}{T_1}$$

Step 3: Substitute the given information into the equation and solve.

$$V_2 = 2.58 \text{ L} \times \frac{341 \text{ K}}{318 \text{ K}} = 2.77 \text{ L}$$

PROBLEM 60

The pressure of a container of helium is 650.8 torr at 35°C. If the sealed container is cooled to 3°C, what will the pressure be?

SOLUTION:

Use the Gay-Lussac's law in the form of equation 65.
Step 1: Organize the given information.

$P_1 = 650.8$ torr $\qquad\qquad$ $T_1 = 35°C + 273 = 308$ K

$P_2 = ?$ $\qquad\qquad$ $T_2 = 3°C + 273 = 276$ K

Step 2: Write and solve equation for the unknown.

$$\frac{P_1}{T_1} = \frac{P_2}{T_2}, \quad \therefore P_2 = P_1 \times \frac{T_2}{T_1}$$

Step 3: Substitute the given information into the equation and solve.

$$P_2 = 650.8 \text{ torr} \times \frac{276 \text{ K}}{308 \text{ K}} = 583 \text{ torr}$$

PROBLEM 61

A gas cylinder contains 40.0 liters of nitrogen gas at 45.5°C and has a pressure of 782.5 torr. What will the pressure be if the temperature is changed to 99.8°C?

SOLUTION:

Use the Gay-Lussac's law in the form of equation 65.
Step 1: Organize the given information.

$P_1 = 782.5$ torr $\qquad\qquad$ $T_1 = 45.5°C + 273 = 318.5$ K

$P_2 = ?$ $\qquad\qquad$ $T_2 = 99.8°C + 273 = 372.8$ K

Step 2: Write and solve equation for the unknown.

$$\frac{P_1}{T_1} = \frac{P_2}{T_2}, \quad \therefore P_2 = P_1 \times \frac{T_2}{T_1}$$

Step 3: Substitute the given information into the equation and solve.

$$P_2 = 782.5 \text{ torr} \times \frac{372.8 \text{ K}}{318.5 \text{ K}} = 915.9 \text{ torr}$$

PROBLEM 62

The temperature of 3.5 liters of helium gas is determined to be 122.5 K at 10.0 atm pressure. What would be the temperature of the gas if the pressure were decreased to 3.2 atm and the volume increased to 6.5 liters?

SOLUTION:

Use the combined gas law in the form of equation 67.

Step 1: Organize the given information.

$$P_1 = 10.0 \text{ atm} \qquad\qquad\qquad P_2 = 3.2 \text{ atm}$$

$$V_1 = 3.5 \text{ L} \qquad\qquad\qquad V_2 = 6.5 \text{ L}$$

$$T_1 = 122.5 \text{ K} \qquad\qquad\qquad T_2 = ?$$

Step 2: Write and solve equation for the unknown.

$$\frac{P_1 V_1}{T_1} = \frac{P_2 V_2}{T_2}, \quad \therefore T_2 = \frac{P_2 V_2 T_1}{P_1 V_1}$$

Step 3: Substitute the given information into the equation and solve.

$$T_2 = \frac{(3.2 \text{ atm})(6.5 \text{ L})(122.5 \text{ K})}{(10.0 \text{ atm})(3.5 \text{ L})} = 72.8 \text{ K}$$

PROBLEM 63

A 5.8 liter flask contains nitrogen gas at a pressure of 637 mm Hg and temperature of 302 K. What would be the volume of the gas at STP?

SOLUTION:

Use the combined gas law in the form of equation 67.

Step 1: Organize the given information.

$$P_1 = 637 \text{ mm Hg} \qquad\qquad P_2 = 760 \text{ mm Hg (value @STP)}$$

$$V_1 = 5.8 \text{ L} \qquad\qquad\qquad\qquad V_2 = ?$$

$$T_1 = 302 \text{ K} \qquad\qquad\qquad T_2 = 273 \text{ K (value @STP)}$$

Step 2: Write and solve equation for the unknown.

$$\frac{P_1 V_1}{T_1} = \frac{P_2 V_2}{T_2}, \quad \therefore V_2 = \frac{P_1 V_1 T_2}{P_2 T_1}$$

Step 3: Substitute the given information into the equation and solve.

$$V_2 = \frac{(637 \text{ mm Hg})(5.8 \text{ L})(273 \text{ K})}{(760 \text{ mm Hg})(302 \text{ K})} = 4.4 \text{ L}$$

PROBLEM 64

The pressure of a 14.0 L sample of H_2 gas at 127°C is determined to be 8.8 atm. If the temperature is decreased to -23°C while the volume is also decreased to 5.6 L, what would be the final pressure of the gas?

SOLUTION:

Use the combined gas law in the form of equation 67.

Step 1: Organize the given information.

$P_1 = 8.8$ atm $P_2 = ?$

$V_1 = 14.0$ L $V_2 = 5.6$ L

$T_1 = 400$ K $T_2 = 250$ K

Step 2: Write and solve equation for the unknown.

$$\frac{P_1 V_1}{T_1} = \frac{P_2 V_2}{T_2}, \quad \therefore P_2 = \frac{P_1 V_1 T_2}{V_2 T_1}$$

Step 3: Substitute the given information into the equation and solve.

$$P_2 = \frac{(8.8 \text{ atm})(14.0 \text{ L})(250 \text{ K})}{(5.6 \text{ L})(400 \text{ K})} = 13.8 \text{ atm}$$

PROBLEM 65

A sample of ammonia gas (NH_3) occupied 6.5 L at 355 torr and was manipulated until its volume was 2.9 L and its final pressure was 196 torr at -138°C. What was the initial temperature of the gas?

SOLUTION:

Use the combined gas law in the form of equation 67.

Step 1: Organize the given information.

$P_1 = 335$ torr $P_2 = 196$ torr

$V_1 = 6.5$ L $V_2 = 2.9$ L

$T_1 = ?$ $T_2 = 135$ K

Step 2: Write and solve equation for the unknown.

$$\frac{P_1 V_1}{T_1} = \frac{P_2 V_2}{T_2}, \quad \therefore T_1 = \frac{P_1 V_1 T_2}{P_2 V_2}$$

Step 3: Substitute the given information into the equation and solve.

$$T_1 = \frac{(335 \text{ torr})(6.5 \text{ L})(135 \text{ K})}{(2.9 \text{ L})(196 \text{ torr})} = 517 \text{ K}$$

PROBLEM 66

A sample of gas originally at -88°C and 2.7 atm occupied a certain volume. When the pressure was decreased to 1.8 atm at 128°C, the volume occupied by the gas was 21.8 L. What was the initial volume of the gas?

SOLUTION:

Use the combined gas law in the form of equation 67.
Step 1: Organize the given information.

$$P_1 = 2.7 \text{ atm} \qquad\qquad P_2 = 1.8 \text{ atm}$$

$$V_1 = ? \qquad\qquad V_2 = 21.8 \text{ L}$$

$$T_1 = 185 \text{ K} \qquad\qquad T_2 = 401 \text{ K}$$

Step 2: Write and solve equation for the unknown.

$$\frac{P_1 V_1}{T_1} = \frac{P_2 V_2}{T_2}, \quad \therefore V_1 = \frac{P_2 V_2 T_1}{P_1 T_2}$$

Step 3: Substitute the given information into the equation and solve.

$$V_1 = \frac{(1.8 \text{ atm})(21.8 \text{ L})(185 \text{ K})}{(2.7 \text{ atm})(401 \text{ K})} = 6.7 \text{ L}$$

PROBLEM 67

A 14.8 L sample of gas contains 0.66 mol of oxygen gas (O_2) at 1 atm pressure at 25°C. The oxygen gas (O_2) was converted to ozone gas (O_3) at the same temperature and pressure. (a) How many moles of ozone is produced. (b) What would be the volume of the ozone?

SOLUTION:

The balanced equation for the reaction is

$$3O_2(g) \rightarrow 2O_3(g)$$

To calculate the moles of O_3 produced, use the appropriate mole ratio:

$$0.66 \text{ mol } O_2 \times \frac{2 \text{ mol } O_3}{3 \text{ mol } O_2} = 0.44 \text{ mol } O_3$$

Avogadro's law expressed as shown in equation 70 states that,

$$\frac{V_1}{n_1} = \frac{V_2}{n_2}$$

If V_1 is the volume of n_1 moles of O_2 gas and V_2 is the volume of n_2 moles of O_3 gas, then

$$n_1 = 0.66 \text{ mol} \qquad\qquad n_2 = 0.44 \text{ mol}$$

$$V_1 = 14.8 \text{ L} \qquad\qquad\qquad V_2 = ?$$

Solving for V_2 gives,

$$V_2 = V_1 \times \frac{n_2}{n_1} = 14.8 \text{ L} \times \frac{0.44 \text{ mol}}{0.66 \text{ mol}} = 9.9 \text{ L}$$

PROBLEM 68

What pressure will be exerted by 0.45 mol of helium gas in a 6.00 L container at 17.0°C?

SOLUTION:

Use the ideal gas law in the form of equation 74.
Step 1: Organize the given information, putting temperature in Kelvin.

$$P = ?$$
$$V = 6.00 \text{ L}$$
$$T = 290.0 \text{ K}$$
$$n = 0.45 \text{ mol}$$
$$R = 0.08206 \text{ L·atm·mol}^{-1}\text{·K}^{-1}$$

Step 2: Write and solve equation for the unknown.

$$PV = nRT, \quad \therefore P = \frac{nRT}{V}$$

Step 3:   Substitute the given information into the equation and calculate.

$$P = \frac{(0.45 \text{ mol})(0.08206 \text{ L} \cdot \text{atm} \cdot \text{K}^{-1} \cdot \text{mol}^{-1})(290.0 \text{ K})}{(6.00 \text{ L})} = 1.78 \text{ atm}$$

PROBLEM 69

How many moles of oxygen gas are in a 25.0-liter tank at 13.0°C if the pressure gauge reads 1000.0 lb/in.^{2}?

SOLUTION:

Use the ideal gas law in the form of equation 74.
Step 1: Organize the given information, changing temperature and pressure to Kelvin and atmosphere units, respectively.

$$P = 1000 \text{ lb / inch}^2 \times \frac{1 \text{ atm}}{14.7 \text{ lb / inch}^2} = 68 \text{ atm}$$

$$V = 25.00 \text{ L}$$
$$T = 286.0 \text{ K}$$
$$n = ?$$
$$R = 0.08206 \text{ L ·atm· mol}^{-1}\text{· K}^{-1}$$

Step 2: Write and solve equation for the unknown.

$$PV = nRT, \quad \therefore n = \frac{PV}{RT}$$

Step 3: Substitute the given information into the equation and calculate.

$$n = \frac{(68 \text{ atm})(25.0 \text{ L})}{(0.08206 \text{ L} \cdot \text{atm} \cdot \text{K}^{-1} \cdot \text{mol}^{-1})(286 \text{ K})} = 72 \text{ mol O}_2$$

PROBLEM 70

What volume would 72.0 grams of carbon dioxide gas (CO_2) occupy at $107°C$ and 800 torr?

SOLUTION:
Use the ideal gas law in the form of equation 74.

$$PV = nRT, \quad \therefore V = \frac{nRT}{P}$$

Substitute the given information into the equation, using appropriate factors to convert grams to mole and torr to atmosphere units.

$$V = \frac{72.00 \text{ g CO}_2 \times \left(\dfrac{1 \text{ mol CO}_2}{44.0 \text{ g CO}_2} \right) \times (0.08206 \text{ L} \cdot \text{atm} \cdot \text{K}^{-1} \cdot \text{mol}^{-1})(379 \text{ K})}{(800 \text{ torr})(1 \text{ atm} / 760 \text{ torr})} = 483 \text{ L}$$

PROBLEM 71

How many grams of sulfur trioxide (SO_3) are in 10.2 liter container if the gas exerts 512 torr at $10°C$?

SOLUTION:
Use the ideal gas law in the form of equation 74.

$$PV = nRT$$

$$\therefore n = \frac{PV}{RT} = \frac{(512 \text{ torr})(1 \text{ atm} / 760 \text{ torr})(10.2 \text{ L})}{(0.08206 \text{ L} \cdot \text{atm} \cdot \text{K}^{-1} \cdot \text{mol}^{-1})(283 \text{ K})} = 0.296 \text{ mol SO}_3$$

$$\text{g SO}_3 = 0.296 \text{ moles SO}_3 (80.1 \text{ g SO}_3/1 \text{ mol SO}_3) = 23.71 \text{ g SO}_3$$

PROBLEM 72

A student collected natural gas from a laboratory jet gas at room temperature ($25°C$) in a 250-ml flask. The pressure and mass of the gas at this temperature was 550 torr and 0.118 g, respectively. From these data, calculate the molecular weight of the gas.

SOLUTION:
Using the ideal gas law, solve for the number of moles, n.

$$\therefore n = \frac{PV}{RT}$$

Tabulate the data: $R = 0.08206$ L atm mol^{-1} K^{-1}.

$$P = 0.724 \text{ atm} \qquad\qquad\qquad T = 298 \text{ K}$$

$$V = 0.250 \text{ L} \qquad\qquad\qquad\qquad n = ?$$

On substituting,

$$n = \frac{(0.724 \text{ atm})(0.250 \text{ L})}{(0.08206 \text{ L} \cdot \text{atm} \cdot \text{K}^{-1} \cdot \text{mol}^{-1})(298 \text{ K})} = 0.00740 \text{ mol gas}$$

Molecular weight = grams/mole:

$$MW = \frac{0.118 \text{ g}}{0.00740 \text{ mol}} = 15.9 \text{ g} / \text{mol}$$

PROBLEM 73

Using ideal gas law, calculate the density (in g/liter) of carbon monoxide gas (CO) at 37°C and 415 torr.

SOLUTION:

Use the ideal gas law in the form of equation 74.
$PV = nRT$; n being the number of moles = g/MW, thus

$$PV = (g/MW)(RT)$$

$$\text{Density } (\rho) = g / \text{vol.} = \frac{P(MW)}{RT}$$

$$\therefore \rho = \frac{(415 \text{ torr})(1 \text{ atm} / 760 \text{ torr})(44.0 \text{ g} / \text{mol CO})}{(0.08206 \text{ L} \cdot \text{atm} \cdot \text{K}^{-1} \cdot \text{mol}^{-1})(310 \text{ K})} = 0.94 \text{ g} / \text{L}$$

PROBLEM 74

A 27.50 liter gas tank contains 2.4 moles of nitrogen (N_2) gas and 2.8 moles oxygen gas (O_2) at 45°C. What is the total pressure exerted by the mixture?

SOLUTION:

Use Dalton's law of partial pressures in the form of equation 79.

$$P_{TOTAL} = \frac{n_1 RT}{V} + \frac{n_2 RT}{V}$$

Applying this to P_{N_2} and P_{O_2}

$$P_{N_2} = \frac{n_{N_2} RT}{V} = \frac{2.4 \text{ mol } (0.08206 \text{ L} \cdot \text{atm} \cdot \text{K}^{-1} \cdot \text{mol}^{-1})(318 \text{ K})}{27.50 \text{ L}} = 2.3 \text{ atm}$$

$$P_{O_2} = \frac{n_{O_2} RT}{V} = \frac{2.8 \text{ mol } (0.08206 \text{ L} \cdot \text{atm} \cdot \text{K}^{-1} \cdot \text{mol}^{-1})(318 \text{ K})}{27.50 \text{ L}} = 2.7 \text{ atm}$$

$$P_{TOTAL} = P_{N_2} + P_{O_2} = 2.3 \text{ atm} + 2.7 \text{ atm} = 5.0 \text{ atm}$$

PROBLEM 75

The total pressure exerted by a mixture of unknown quantity of nitrogen and 3.4 moles of oxygen in a 13.5 liter container at 47°C is 12.4 atm. What is the pressure exerted by the nitrogen?

SOLUTION:

Use Dalton's law of partial pressures in the form of equation 79.
Convert 47°C to °Kelvin using equation 60; 47°C = 320 K.
Then,

$$P_{O_2} = \frac{n_{O_2} RT}{V} = \frac{3.4 \text{ mol } (0.08206 \text{ L} \cdot \text{atm} \cdot \text{K}^{-1} \cdot \text{mol}^{-1})(320 \text{ K})}{13.50 \text{ L}} = 6.6 \text{ atm}$$

$$P_{TOTAL} = P_{N_2} + P_{O_2}$$

$$\therefore P_{N_2} = P_{TOTAL} - P_{O_2} = 12.4 \text{ atm} - 6.6 \text{ atm} = 5.8 \text{ atm}$$

PROBLEM 76

A 240.0 mL sample of gas was collected over water at 70°C and was found to exert a pressure of 752 torr. What pressure would the sample exert if dried over calcium chloride without changing the temperature?

SOLUTION:

Use the principles of partial pressure and Boyle's law.
Determine the pressure of the dry gas

$$P_{H_2O} \text{ at } 70°C = 233.7 \text{ torr (Table L1, Appendix K)}$$

$$P_{TOTAL} = P_{gas} + P_{water}$$

$$P_{gas} = P_{TOTAL} - P_{water} = 752 \text{ torr} - 233.7 \text{ torr} = 518.3 \text{ torr}$$

PROBLEM 77

A chemist reacted 0.170 g hydrogen gas (H_2) and 1.22 g of carbon monoxide gas (CO) in a 3.00 L reaction vessel at 298 K. (a) Determine the total pressure of the mixture and (b) find the partial pressures and mole fractions of the two gases.

SOLUTION:

Step 1: Organize the given information.

$$V = 3.00 \text{ L}, T = 298 \text{ K}, \text{mass}_{H_2} = 0.170 \text{ g}, \text{mass}_{CO} = 1.22 \text{ g}$$

Step 2: Convert the mass into moles (n) for each gas.

$$n = \text{mass/MW}; \therefore n_{H_2} = 8.4 \times 10^{-2} \text{ mol and } n_{CO} = 4.4 \times 10^{-2} \text{ mol}$$

Step 3: **Use Dalton's law (equation 79) to compute the partial pressure of each gas.**

$$P_{CO} = \frac{n_{CO}RT}{V} = \frac{4.4 \times 10^{-2} \text{ mol } (8.206 \times 10^{-2} \text{ L} \cdot \text{atm} \cdot \text{K}^{-1} \cdot \text{mol}^{-1})(298 \text{ K})}{3.00 \text{ L}} = 0.359 \text{ atm}$$

The same calculation for H_2 gives $P_{H_2} = 0.685$ atm.

The total pressure is the sum of the partial pressures.

$$P_{total} = P_{CO} + P_{H_2} = (0.359 \text{ atm} + 0.685 \text{ atm}) = 1.044 \text{ atm}$$

Mole fractions can be calculated from moles or partial pressures:

$$\chi_{CO} = \frac{n_{CO}}{n_{total}} = \frac{4.4 \times 10^{-2} \text{ mol}}{12.8 \times 10^{-2} \text{ mol}} = 0.344$$

$$\chi_{H_2} = \frac{P_{H_2}}{P_{total}} = \frac{0.685 \text{ atm}}{1.044 \text{ atm}} = 0.656$$

To check for reasonableness, add the mole fractions, which must sum to 1.00.

PROBLEM 78

What is the molar mass of urea, if a solution was prepared by adding 20.00 g urea to 125.00 g of water at 25°C. Water has vapor pressure of 23.76 torr at this temperature. The observed vapor pressure of the solution was found to be 22.70 torr.

SOLUTION:

Using Raoult's law,

$$P_{solution} = \chi_{solvent} \, P^{o}_{solvent}$$

we write

$$\chi_{H_2O} = \frac{P_{solution}}{P_{H_2O}} = \frac{22.70 \text{ torr}}{23.76 \text{ torr}} = 0.955$$

To find the molar mass of urea, first find the number of moles represented by 20.0 g. Calculate this as follows:

$$\chi_{H_2O} = \frac{\text{mol H}_2\text{O}}{\text{mol H}_2\text{O} + \text{mol urea}} = \frac{n_{H_2O}}{n_{H_2O} + n_{urea}}$$

Since the mass of water is given, then

$$n_{H_2O} = 125.0 \text{ g H}_2\text{O} \times \frac{1 \text{ mol H}_2\text{O}}{18.0 \text{ g H}_2\text{O}} = 6.94 \text{ mol H}_2\text{O}$$

and since $\chi_{H_2O} = 0.955$, we deduce that

$$\chi_{H_2O} = 0.955 = \frac{n_{H_2O}}{n_{H_2O} + n_{urea}} = \frac{6.94}{6.94 + n_{urea}}$$

i.e. $0.955 (6.94 + n_{urea}) = 6.94$, and

$$n_{urea} = \frac{6.94 - 6.62}{0.955} = 0.335 \text{ mol}$$

Since 20.0 g of urea was originally dissolved, 0.335 mol urea weighs 20.0 g. From equation 1,

$$\text{mol} = \frac{g}{\text{molar mass}}$$

$$\therefore \text{molar mass} = \frac{g}{\text{mol}} = \frac{20.0 \text{ g}}{0.335 \text{ mol}} = 59.7 \text{ g / mol}$$

(Urea has a formula of $(NH_2)_2CO$ with molar mass of 60.0 g/mol; our answer from the calculation agrees well to that from the formula).

PROBLEM 79

What is the boiling point elevation of a solution prepared by adding 43.22 grams of urea (molar mass = 60.0 g/mol) to 160.0 mL of water?

SOLUTION:

Use equations 12 and 82.

$$\Delta t_b = K_b m_{solute}$$

$$m_{solute} = \text{moles solute/kg solvent} = \text{moles urea/kg water}$$

$$m_{solute} = 43.22 \text{ g urea} \times \frac{1 \text{ mol urea}}{60.0 \text{ g urea}} = 0.720 \text{ mol urea}$$

$$\text{g solvent} = 160.0 \text{ g } H_2O \times \frac{1 \text{ Kg } H_2O}{1000 \text{ g } H_2O} = 0.160 \text{ Kg } H_2$$

$$\Delta T_b = 0.52°C / m \times \frac{0.720 \text{ mol urea}}{0.160 \text{ Kg } H_2O} = 2.3°C$$

$$T_b = 100°C + \Delta T_b = 100°C + 2.3°C = 102.3°C$$

PROBLEM 80

How many grams of urea (molar mass = 60.0) will be added to 500.0 mL of water in order to prepare a solution with a boiling point 102.7°C?

SOLUTION:

Use equations 12 and 82.

$$\Delta T_b = 102.7°C - 100°C = 2.7°C$$

$$\Delta T_b = K_b m_{solute} = \frac{K_b \ (g / MW)}{kg}$$

$$\therefore \ g_{urea} = \frac{\Delta T_b \cdot kg \cdot MW}{K_b}$$

Substituting the appropriate values,

$$g_{urea} = (2.7°C)\left(500.0 \ mL \times \frac{1 \ g}{1 \ mL} \times \frac{1 \ kg}{1000 \ g} \right) \times (60.0 \ g / mol)\left(\frac{1}{0.52 °C \ kg / mol} \right)$$

$$= 155.77 \ g$$

PROBLEM 81

When 8.04 grams of an unknown compound was dissolved in 26.90 grams of water, a solution with a boiling point of 101.3°C was produced. What is the molecular weight of the unknown?

SOLUTION:

Use equations 12 and 82.

$$\Delta T_b = T_b - 100°C = 101.3°C - 100°C = 1.3°C$$

$$\Delta T_b = K_b m_{solute} = \frac{K_b \ (g / MW)}{kg}$$

$$\text{Thus, MW} = \frac{(K_b)(g_{solute})}{(K_{g_{solute}})(\Delta T_b)}$$

Substituting the appropriate values,

$$MW = \frac{(0.52°C \ kg / mol)(8.04 \ g)}{(26.9 \ g)(1 \ Kg / 1000 \ g)(1.3°C)} = 119.55 \ g / mol$$

PROBLEM 82

A solution is made by dissolving 117.88 g of ethylene glycol ($C_2H_6O_2$) in 204.00 g of water. What is the freezing point of the solution?

SOLUTION:

Make use of equation 83.

To calculate the freezing point of the solution, first calculate ΔT_f using equation 83.

$$\Delta T_f = K_f m_{solute} = K_f \times \frac{mol \ solute}{kg \ solvent}$$

K_f (for water) is 1.86°C kg solvent/mol solute (Table L1).

$$\text{mol solute} = 117.88 \text{ g } C_2H_6O_2 \times \frac{1 \text{ mol } C_2H_6O_2}{62.1 \text{ g } C_2H_6O_2} = 1.90 \text{ mol } C_2H_6O_2$$

$$\text{Kg solvent} = 204.00 \text{ g } H_2O \times \frac{1 \text{ Kg } H_2O}{1000 \text{ g } H_2O} = 0.204 \text{ Kg } H_2O$$

$$\therefore \quad \Delta T = \frac{1.86°C \text{ Kg } H_2O}{0.204 \text{ Kg } H_2O} \times \frac{1.90 \text{ mol } C_2H_6O_2}{1 \text{ mol } C_2H_6O_2} = 17.3°C$$

The freezing point of the solution = freezing point of solvent - ΔT_f

$$= 0.0°C - 17.3°C$$

$\therefore$ The freezing point of the solution = - 17.3°C

PROBLEM 83

A solution made by dissolving 4.72 g of an unknown compound in 100.00 g of water has a freezing point of -1.46°C. What is the molecular weight of the unknown?

SOLUTION:
Make use of equations 12 and 83.

$$\Delta T_f = 0°C - T_f$$
$$= 0°C - (-1.46) = 1.46°C$$
$$\Delta T_f = K_f m_{solute}$$
$$\Delta T_f = K_f(g/MW/kg)$$
$$(MW)(kg)(\Delta T_f) = (K_f)(g)$$

$$MW = \frac{(K_f)(g_{solute})}{(\Delta T_f)(Kg_{solute})} = \frac{(1.86°C / mol)(4.72 \text{ g})}{(1.46°C)(100.00 \text{ g})(1 \text{ Kg} / 1000.00 \text{ g})} = 60.1 \text{ g / mol}$$

PROBLEM 84

Calculate the osmotic pressure at 37°C of a solution prepared by adding 18.70 g of glucose ($C_6H_{12}O_6$, MW = 180.0 g/mol) to 49.30 g of water. The density of the solution is 1.06 g/ml.

SOLUTION:
Use equation 84; $\Pi = MRT$, where Π is the osmotic pressure.

$$g_{soln} = 18.7 \text{ g} + 49.3 \text{ g} = 68.0 \text{ g}$$
$$V_{soln} = 68.0 \text{ g} (1 \text{ ml}/1.06 \text{ g}) = 64.2 \text{ mL}$$

$$M = \frac{18.7 \text{ g}(1 \text{ mol} / 180.0 \text{ g})}{(64.2 \text{ ml})(1 \text{ L} / 1000 \text{ ml})} = 1.62 \text{ mol / L}$$

$$T = 273 + 37°C = 310 \text{ K}, \qquad R = 0.08206 \text{ L·atm/mol·K}$$

$$\Pi = (1.62 \text{ mol/L})(0.08206 \text{ L·atm/mol·K})(310 \text{ K}) = 41.2 \text{ atm}$$

PROBLEM 85

How many grams of glucose (MW = 180.0 g/mol) will be needed in order to make 240.0 mL of a solution with osmotic pressure of 2.9 atm at 58°C?

SOLUTION:

Using equation 84 again we have,

$$\Pi = MRT$$

$$\text{Thus, } M = \frac{\Pi}{RT} = \frac{2.9 \text{ atm}}{(0.08206 \text{ L} \cdot \text{atm} \cdot \text{K}^{-1} \cdot \text{mol}^{-1})(331.0 \text{ K})} = 0.107 \text{ mol} / \text{L}$$

$$g = M \times L \times MW \text{ (equation 9)}$$

$$= (0.107 \text{ mol/L})(0.240)(180.0 \text{ g/mol})$$

$$= 4.62 \text{ grams}$$

PROBLEM 86

A 50.0 ml aqueous solution containing 0.51 g of hemoglobin has an osmotic pressure of 4.8 torr at 37°C. What is the molecular weight of the hemoglobin?

SOLUTION:

We use equation 84 again.

$$\Pi = MRT$$

We can also rewrite this equation as

$$\Pi = \frac{mRT}{V(MW)}$$

Given:

$$m_{\text{solute}} = 0.51 \text{ g} \quad T = 310 \text{ K} \quad R = 8.206 \times 10^{-2} \text{ L} \cdot \text{atm} \cdot \text{K}^{-1} \cdot \text{mol}^{-1}$$

$$\Pi = (4.8 \text{ torr})(1 \text{ atm} /760 \text{ torr}) = 6.32 \times 10^{-3} \text{ atm}$$

$$V_{\text{solution}} = (50.0 \text{ mL})(1 \text{ L}/1000 \text{ mL}) = 5.0 \times 10^{-2} \text{ L}$$

$$MW = \frac{mRT}{\Pi V} = \frac{(0.51 \text{ g})(8.206 \times 10^{-2} \text{ L} \cdot \text{atm} \cdot \text{K}^{-1} \cdot \text{mol}^{-1})(310 \text{ K})}{(6.32 \times 10^{-3} \text{ atm})(5.0 \times 10^{-2} \text{ L})} = 4.11 \times 10^{4} \text{ g} / \text{mol}$$

PROBLEM 87

What is the vapor pressure of a solution prepared by adding 45.05 g of glucose ($C_6H_{12}O_6$) to 150.0 ml of water at 60°C?

SOLUTION:

Use equation 85.

$$P_{\text{solution}} = \chi_{\text{water}} P^{\circ}_{\text{water}}$$

$$\text{mol glucose} = (45.05 \text{ g}) \times \frac{1 \text{ mol}}{180.0 \text{ g}} = 0.250 \text{ mol}$$

$$\text{mol } H_2O = 150.0 \text{ mL } (1 \text{ g }/1 \text{ mL})(1 \text{ mol}/18.0 \text{ g}) = 8.33 \text{ mol}$$

$$\chi_{H_2O} = \frac{\text{mol } H_2O}{\text{total mol}} = \frac{8.33 \text{ mol}}{(0.250 + 8.33) \text{ mol}} = 0.971$$

$$P^{\circ}_{H_2O} \text{ at } 60°C = 149.4 \text{ torr (see Table L1)}$$

$$P_{\text{solution}} = (0.971)(149.4) = 145.1 \text{ torr}$$

Chapter 4

Buffers: Principles,
Calculations, and Preparation

4.1. Principles

A. Definitions and Abbreviations

A pH buffer usually consists of a weak acid and its conjugate base. The function of pH buffers is to neutralize small changes in H^+ and OH^- concentrations, thereby maintaining a fairly constant pH. The basics of pH buffering, buffer calculations and preparation are discussed below. The necessary symbols and their meanings are as follows: A weak acid and its conjugate base are represented by HA and A^-, respectively; a weak base and its conjugate acid are represented by $R-NH_2$ and $R-NH_3^+$, respectively, because the most commonly used weak base buffers have NH_2 basic functional group(s). In aqueous solutions, H^+ is hydrated and, therefore, is written as H_3O^+. Molar concentration is indicated by enclosing the substance of interest in square brackets; for example, $[H^+]$ represents the molar concentration of H^+.

B. Properties of Acids and Bases

For the purpose of this chapter, acids and bases are defined according to Bronsted-Lowry (*2, 3*). An acid is a chemical compound that donates a H^+, and a base is a compound that accepts a H^+. When an acid is dissolved in H_2O, it donates its H^+ to a H_2O molecule which, in this case, acts as a base. When a base is dissolved, it accepts a H^+ from a H_2O molecule which, in this case, acts as an acid. These reactions are shown below using the symbols HA and $R-NH_2$ for the acid and base, respectively:

$$HA + H_2O \rightleftharpoons H_3O^+ + A^-$$

$$R-NH_2 + H_2O \rightleftharpoons R-NH_3^+ + OH^-$$

In the first equation, the ionized acid (A^-) is termed the conjugate base of HA, and HA is the conjugate acid of A^-. In the second equation, the ionized base ($R-NH_3^+$) is the conjugate acid of $R-NH_2$, and $R-NH_2$ is the conjugate base of $R-NH_3^+$.

As indicated by the double arrows in the equations above, acid-base reactions are reversible, but the extent of reversibility depends on the strength of the acid or base. Strong acids or bases dissociate almost completely in H_2O. For example, when the strong acid HCl is dissolved in H_2O, almost 100% of it dissociates as follows:

$$HCl + H_2O \rightleftharpoons H_3O^+ + Cl^-$$

where the longer upper arrow indicates a strong preference toward formation of the H_3O^+ and Cl^-. In a dilute solution at equilibrium, very little of the HCl remains, and H_3O^+ and Cl^- are the predominant species. The Cl^-, which is the conjugate base of HCl, is a weak base. Similarly, when a strong base is dissolved, the resultant conjugate acid is typically a weak one. **In general, a strong acid produces a weak conjugate base, and a strong base produces a weak conjugate acid.** As explained in the next section, this tendency toward complete ionization, and the production of weak conjugate forms, makes strong acids and bases unsuitable for pH buffering. In contrast to the above properties, a weak acid or base dissociates partially in H_2O. As such, significant quantities of both the conjugate acid and the conjugate base remain at equilibrium. Furthermore, **a weak acid produces a strong conjugate base, and a weak base produces a strong conjugate acid.** These properties make them suitable for pH buffering, the mechanism of which is discussed below.

C. Mechanism of pH Buffering by Weak Acids and Bases

As mentioned above, weak acids and weak bases ionize incompletely when dissolved in H_2O. The solution, therefore, contains significant concentrations of the conjugate acid HA or $R-NH_3^+$ in equilibrium with H_3O^+ and the conjugate base (A^- or $R-NH_2$), as shown in equations 86–90, below.

For a weak acid with one ionizable H,

$$HA + H_2O \rightleftharpoons H_3O^+ + A^- \tag{86}$$

$$K = \frac{[H_3O^+][A^-]}{[H_2O][HA]} \tag{87}$$

where K is the equilibrium constant. If the ionization of H_2O is ignored, equation 86 becomes

$$HA \rightleftharpoons H^+ + A^- \tag{88}$$

$$K_a = \frac{[H^+][A^-]}{[HA]} \tag{89}$$

For a weak base that accepts one H^+,

$$R-NH_2 + H_2O \rightleftharpoons R-NH_3^+ + OH^- \tag{90}$$

$$K_b = \frac{[R-NH_3^+][OH^-]}{[R-NH_2][H_2O]} \tag{91}$$

where K_a and K_b are, respectively, the expression of the equilibrium constant for the acid and base represented above.

When a small quantity of H^+ is added to a solution of a weak acid or base, most of the H^+ released, reacts with the conjugate base (A^- or $R-NH_2$) to form the corresponding conjugate acid (HA or $R-NH_3^+$). If OH^- is added, most of it reacts with the conjugate acid to form the conjugate base and H_2O. Thus, most of the added H^+ or OH^- are not free to appreciably change the total H^+ concentration in the solution and, as such, there is little or no change in the pH of the solution. The weak acid or base is, therefore, a pH buffer since it guards

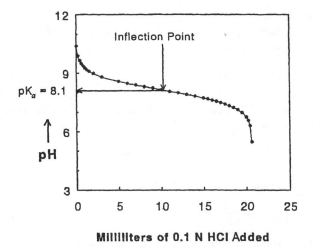

Figure 4.1. The pH titration curve for the monoprotic weak base, tris(hydroxymethyl)aminomethane (Tris). Twenty-five mL of 0.1 M Tris was titrated with small volumes of 0.1 M HCl. The pH of the solution was plotted against the volume of HCl. At the inflection point of the curve, the concentrations of the conjugate base and the conjugate acid are equal, and pH buffering is maximal. This pH (8.1) is the pK_a of Tris.

against changes in the pH of the solution. The resistance to pH changes is greatest at the pH where the concentrations of the conjugate acid and conjugate base are equal. This pH is known as the pK_a of the buffer (see Figure 4.1). The effective buffering range is about p$K_a \pm$ 1 pH unit (Figure 4.1), and for this reason, a buffer chosen for a biochemical assay should have a pK_a near the optimum pH for the assay.

A strong acid or base (e.g., HCl or KOH) cannot buffer pH because it dissociates almost completely in dilute aqueous solutions, and the conjugate base or acid is weak. If, for example, a small amount of H^+ is added to a dilute solution of HCl, the total H^+ concentration will increase because the conjugate base Cl^- is too weak to react with the added H^+. If OH^- is added, it reacts with free H^+, and since the concentration of HCl in the solution is too low to ionize and replace the reacting H^+, the concentration of H^+ in the solution decreases. In either case, there is no pH buffering.

D. The Ionization of Water

When equations 86 and 87 were written, we ignored the ionization of water. Now lets go back and look at the ionization of water, shown below:

$$H_2O + H_2O \rightleftharpoons H_3O^+(aq) + OH^-(aq) \qquad (92)$$

A reaction, in which two molecules of solvent react with each other to form ions is called auto-ionization. Since autoionization is an equilibrium, we can write an equilibrium expression for autoionization of water as (see also Appendix A):

$$K_c = \frac{[H_3O^+][HO^-]}{[H_2O][H_2O]} \qquad (93)$$

This is a very important equilibrium because it is present in any aqueous solution, regardless of what other reactions may be taking place. The molar concentration of water which appears in the denominator of the expression above, is nearly constant (≈ 55.6 M) in both dilute aqueous solutions and pure water. Hence, we can include $[H_2O]^2$ in the equilibrium constant K_c of equation 93 above to give,

$$K_c \cdot [H_2O]^2 \rightleftharpoons [H_3O^+][OH^-] \tag{94}$$

The left side of equation 94 is the product of two constants and is also equal a constant. The combined constant can be written as

$$K_w = K_c \cdot [H_2O]^2$$

The equilibrium law then becomes

$$K_w = [H_3O^+][OH^-] \tag{95}$$

E.   The Relationship Between K_a and K_b for an Acid-Base Conjugate Pair

Lets consider the equilibrium constant expression for the ammonium ion-ammonia conjugate pair in water, shown by equations 101 and 103.
For NH_4^+:

$$NH_4^+ + H_2O \rightleftharpoons NH_3 + H_3O^+ \tag{96}$$

$$K_c = \frac{[NH_3][H_3O^+]}{[NH_4^+][H_2O]} \tag{97}$$

$$K_c [H_2O] = \frac{[NH_3][H_3O^+]}{[NH_4^+]} \tag{98}$$

$$K_a = \frac{[NH_3][H_3O^+]}{[NH_4^+]} \tag{99}$$

The simplified equation for the equilibrium and the corresponding equilibrium law are

$$NH_4^+ \rightleftharpoons NH_3 + H^+ \tag{100}$$

$$K_a = \frac{[NH_3][H^+]}{[NH_4^+]} \tag{101}$$

For NH_3:

$$NH_3 + H_2O \rightleftharpoons NH_4^+ + HO^-$$

$$K_c = \frac{[NH_4^+][HO^-]}{[H_2O][NH_3]} \tag{102}$$

$$K_c[H_2O] = K_b = \frac{[NH_4^+][HO^-]}{[NH_3]}$$

$$K_b = \frac{[NH_4^+][HO^-]}{[NH_3]} \qquad (103)$$

A simple relationship exits between the values of K_a and K_b for an acid-base conjugate pair, which is $K_a \times K_b = K_w$.

So considering the pair NH_4^+ and NH_3

$$K_a = \frac{[NH_3][H^+]}{[NH_4^+]}$$

$$K_b = \frac{[NH_4^+][HO^-]}{[NH_3]}$$

$$K_a \times K_b = \frac{[NH_3][H^+]}{[NH_4^+]} \times \frac{[NH_4^+][HO^-]}{[NH_3]} = [H^+][HO^-]$$

$$\therefore K_a \times K_b = \mathbf{K_w} \qquad (104)$$

4.2. Buffer, pH, and pOH Calculations

A. The pH/pOH Scale

The solvent for biological pH buffers is H_2O, and for this reason, the pH/pOH scale is based on the ionization constant of pure H_2O. Pure H_2O ionizes slightly as shown before in equation 92.

$$2H_2O \rightleftharpoons H_3O^+ + OH^-$$

or simply,

$$H_2O \rightleftharpoons H^+ + OH^- \qquad (105)$$

The ionization constant, K_w (derived in equation 95 above and in Appendix A.1), can also be written as

$$K_w = [H^+][OH^-] \qquad (106)$$

In pure H_2O at room temperature,

$$[H^+] = [OH^-] = 1 \times 10^{-7} \, M \qquad (107)$$

$$\therefore [H^+][OH^-] = [1 \times 10^{-7}]^2 = 1 \times 10^{-14} \qquad (108)$$

Water or water solution in which $[H^+] = [OH^-] = 1 \times 10^{-7} \, M$ are neutral solutions. A solution in which $[H^+]$ is greater than $[OH^-]$ is acidic, a solution in which $[OH^-]$ is greater than $[H^+]$ is basic. Equation 108 therefore defines the range for the acidity or basicity of aqueous solutions. This range can vary over extreme values, from 1 M or greater to 1×10^{-14} M or less. Data with such a wide range of variations are difficult to handle on a linear scale and are therefore expressed as logarithms to compress the range and ease data handling. Thus, the concentrations of H^+ and OH^- in aqueous solutions are normally expressed as logarithms: pH and pOH, respectively. The scale for pH/pOH is derived by taking the logarithm of equation 108:

$$\log [H^+][OH^-] = \log [1 \times 10^{-14}]$$

or

$$\log [H^+] + \log [OH^-] = -14 \qquad (109)$$

Multiplying both sides of equation 109 by -1 yields

$$-\log [H^+] + -\log [OH^-] = 14 \qquad (110)$$

In general, $-\log X$ is defined as pX.

$$\therefore -\log [H^+] = pH \qquad (111)$$

and

$$-\log [OH^-] = pOH \qquad (112)$$

Substituting equations 111 and 112 into equation 110 yields

$$pH + pOH = 14 \qquad (113)$$

which defines the pH/pOH scale for aqueous solutions. The negative sign in equations 111 and 112, which was arbitrarily introduced by multiplying equation 109 by -1, is intended to make the pH and the pOH values positive since most $[H^+]$ and $[OH^-]$ used in the laboratory are less than 1 M and would otherwise yield negative pH or pOH values. By this convention, a negative pH or pOH will result when $[H^+]$ or $[OH^-]$ is greater than 1 M.

Not all ions of a substance in solution are active; hence, the effective concentration or **activity** of an ion is less than the molar concentration. For this reason, pH and pOH are more accurately defined as:

$$pH = -\log \gamma [H^+] \text{ and } pOH = -\log \gamma[OH^-] \qquad (114)$$

where γ is the activity coefficient of the respective ions. In dilute solutions (0.1 M or less), $\gamma \approx 1$. Therefore, $\gamma[H^+]$ and $\gamma[OH^-] \approx [H^+]$ and $[OH^-]$, respectively, and equations 111 and 112 may be used as written. A pH electrode measures the activity of H^+, i.e., $\gamma[H^+]$.

Equations 111–114 are used to calculate the pH or pOH of an aqueous solution provided that the free $[H^+]$ or $[OH^-]$ in the solution is known.

Examples: See problems 91 and 92.

B.   The Henderson-Hasselbach Equation

The Henderson-Hasselbach equation, derived in Appendix A.2, is shown below. For a weak acid,

$$pH = pK_a + \log \frac{[A^-]}{[HA]} \tag{115}$$

For a weak base,

$$pH = pK_a + \log \frac{[R-NH_2]}{[R-NH_3^+]} \tag{116}$$

The equation relates the pH of a buffer to the pK_a and concentrations of the conjugate acid and conjugate base, and predicts that the pH of a buffer is determined by the (conjugate acid)/(conjugate base) ratio. In other words, a given pH for a buffer is attained once the (conjugate base)/(conjugate acid) ratio predicted for that pH by this equation is established. This implication underlies the methods by which pH buffers are prepared (see below). When $[A^-] = [HA]$, the ratio $[A^-]/[HA]$ becomes 1, log 1 = 0, and equation 115 becomes, pH $= pK_a$. This means that pK_a is the pH at which the concentrations of the conjugate acid and conjugate base are equal.

Buffer calculations often require calculating the $[A^-]/[HA]$ ratio. An expression for a direct calculation of this ratio is derived by solving equation 115 as follows:

$$pH = pK_a + \log \frac{[A^-]}{[HA]}$$

$$\log \frac{[A^-]}{[HA]} = pH - pK_a \tag{117}$$

Upon taking the antilog,

$$\frac{[A^-]}{[HA]} = 10^{(pH-pK_a)} \tag{118}$$

For a weak base, equation 118 is similarly solved to yield:

$$\frac{[R-NH_2]}{[R-NH_3^+]} = 10^{(pH-pK_a)} \tag{119}$$

Equations 115–119, in conjunction with equation 9, are used to calculate the amounts of the conjugate acid and conjugate base which, when dissolved, yield a buffer of a given pH, concentration, and volume.

Examples: See problems 88 and 89.

C. Buffer Capacity (BC)

Buffer capacity describes the ability of a buffer to resist changes in pH when H^+ or OH^- is added. For quantitative purposes, it can be defined as the number of mol/L of H^+ or OH^- needed to change the pH of a buffer by 1 unit. In other words, BC is the change in the total $[H^+]$ or $[OH^-]$ per unit change in pH:

$$BC = \frac{d[H^+]_T}{d\text{pH}} = -\frac{d[OH^-]_T}{d\text{pH}} \qquad (120)$$

where $d[H^+]_T$ is the change in the total H^+ concentration (i.e., bound plus free $[H^+]$) and is given by equation 126, below. To calculate BC, $d[H^+]_T$ is substituted by equation 126. When this substitution is made, equation 120 reveals that the capacity of any buffer is dependent on its concentration, maximal at the pK_a, and drops off to lower values as pH deviates from the pK_a. BC values outside the range pH = $pK_a \pm 1$ are too low for effective buffering, and buffers should not be used at pH values outside this range.

Often, a reaction in a buffered assay medium releases or takes up H^+. The total amount of H^+ released or taken up is represented by $d[H^+]_T$ and can be calculated using the Henderson-Hasselbach equation, or directly from equation 126 which is derived as follows: When a small amount of H^+ is added to a buffer, most of the ions bind to the conjugate base (A^-) to form the conjugate acid (HA), and a small amount remains as free H^+. Therefore, [HA] and $[H^+]$ in the buffer have changed by the amounts $\Delta[HA]$ and $\Delta[H^+]$, respectively. This means that

$$d[H^+]_T = \Delta[HA] + \Delta[H^+]$$

Let the initial pH of the buffer be pH_1 and the final pH be pH_2. Then,

$$D[H+]_T = ([HA_{pH_2}] - [HA_{pH_1}]) + ([H^+_{pH_2}] - [H^+_{pH_1}]) \qquad (121)$$

From equation 118,

$$\frac{[A^-]}{[HA]} = 10^{(\text{pH}-pK_a)}$$

$$\therefore [A^-] = [HA] \times 10^{(\text{pH}-pK_a)} \qquad (122)$$

Also,

$$[HA] + [A^-] = C \qquad (123)$$

where C is the total molarity of the buffer. Substituting equation 122 into equation 123 yields:

$$[HA] + \left([HA] \times 10^{(\text{pH}-pK_a)}\right) = C$$

$$[HA]\left(1 + 10^{(\text{pH}-pK_a)}\right) = C$$

$$\therefore \quad [HA] = \frac{C}{1 + 10^{(\text{pH}-pK_a)}} \qquad (124)$$

From equation 111,

$$\text{pH} = -\log[H^+]$$

$$\therefore [H^+] = 10^{-\text{pH}} \qquad (125)$$

Substituting equations 124 and 125 into equation 121 yields:

$$d[\text{H}^+]_\text{T} = \left(\frac{C}{1 + 10^{(\text{pH}_2 - \text{p}K_a)}} - \frac{C}{1 + 10^{(\text{pH}_1 - \text{p}K_a)}} \right) + (10^{-\text{pH}_2} - 10^{-\text{pH}_1}) \quad (126)$$

Example: See problem 90.

4.3. Preparation of Buffers and Related Topics

A. Preparation of Buffers

Methods used to prepare buffers are based on the basic principle specified by the Henderson-Hasselbach equation: That the pH of a buffer is determined by the (conjugate base)/(conjugate acid) ratio, and vice versa. Preparing a buffer of a given pH, therefore, requires the preparation of a solution that contains the conjugate base and the conjugate acid at a ratio equal to the value predicted by the Henderson-Hasselbach equation. The appropriate ratio can be established by adding acid or base to a solution of one of the conjugates to adjust the pH to the desired value; or by dissolving precalculated quantities of the conjugates. The following three methods are based on these principles.

Method 1. Simple pH adjustment

(1) Calculate the weight of the buffer needed to prepare a solution of a given volume and molarity using equation 9 (use any of the conjugate forms of the buffer).
(2) Dissolve it in a volume of deionized H_2O, which is about 80% of the desired final volume.
(3) Adjust the pH to the desired value, then add more H_2O to adjust the volume to the final value.

Method 2. Using calculated amounts of the conjugate forms

(1) Use the Henderson-Hasselbach equation to calculate the concentrations of the conjugate acid and base that will be present in the buffer at the given pH.
(2) Use equation 9 to calculate the weight corresponding to each of the concentrations in step 1.
(3) Dissolve these quantities as in step 2 of method 1, above.
(4) Adjust the volume to the target value by adding additional H_2O.
(5) Check the pH. It should be within ±0.2 units of the expected value.

Examples: See problems 88a–d.

Method 3. Mixing proportionate volumes of the conjugate forms

(1) For each conjugate, calculate the weight needed to prepare a solution at a concentration equal to that of the buffer.
(2) Calculate the ratio $[\text{A}^-]/[\text{HA}]$ or $[\text{R-NH}_2]/[\text{R-NH}_3^+]$ at the given pH of the buffer. Use equations 115 or 116 (or equations 118 or 119).
(3) Use this ratio to calculate the volumes of both solutions which, when combined, yield a buffer with the given pH, concentration, and volume.

(4) Prepare the buffer by combining the measured volumes. Check the pH. It should be within ± 0.2 units of the expected value.

Examples: See problems 89a–c.

Recipes for preparing common laboratory buffers based on Method 3 are given in Appendix H.

B.   Changes in Buffer pH upon Dilution

When a concentrated buffer is diluted to prepare a working buffer, the pH of the dilute buffer changes even though no H^+ or OH^- has been added. The reasons for this change are manifold, but only the most important two are discussed below.

(1)   Effect of dilution on activity coefficient

Some of the conjugate species in a buffer are ions. The effective molar concentration or activity of dissolved ions is defined as follows:

$$\text{Activity} = \gamma M \tag{127}$$

where γ and M are the activity coefficient and molarity of the ion, respectively. Substituting this into the Henderson-Hasselbach equation yields:
For a weak acid,

$$pH = pK_a + \log \frac{\gamma[A^-]}{[HA]} \tag{128}$$

For a weak base,

$$pH = pK_a + \log \frac{[R\text{-}NH_2]}{\gamma[R\text{-}NH_3^+]} \tag{129}$$

Because activity coefficients increase with increasing dilution, γ [A^-]/[HA] or [R-NH$_2$]/γ[R-NH$_3^+$] will change upon dilution, resulting in a change in the buffer's pH. For a weak acid, the pH increases because log (γ[A^-]/[HA]) increases, and for a weak base, the pH decreases because log ([R-NH$_2$]/γ[R-NH$_3^+$]) decreases with increasing dilution.

(2)   Effect of dilution on the ionization of acids and bases

The degree of ionization of acids and bases increases as their concentration decreases with dilution. Thus, [A^-] or [R-NH$_2$] increases while [HA] or [R-NH$_3^+$] decreases. Log ([A^-]/[HA]) or log [RNH$_2$]/[R-NH$_3^+$] increases accordingly, resulting in a pH increase upon dilution of a weak acid or a weak base buffer.

C.   Polyprotic Weak Acids

Polyprotic acids have more than one ionizable hydrogen. When dissolved, the hydrogen atoms dissociate sequentially as the pH of the solution increases, resulting in a stepwise profile in a pH titration curve (Figure 4.2). To illustrate this, the triprotic acid H_3PO_4 ionizes in three steps as follows:

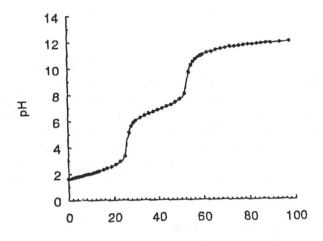

Figure 4.2. The pH titration curve for the triprotic weak acid, H_3PO_4.

$$H_3PO_4 \overset{K_{a_1}}{\rightleftharpoons} H^+ + H_2PO_4^- \overset{K_{a_2}}{\rightleftharpoons} H^+ + HPO_4^{2-} \overset{K_{a_3}}{\rightleftharpoons} H^+ + PO_4^{3-}$$

The ionization equilibrium can be written in three steps as though three different weak acids were involved.

$$H_3PO_4 \rightleftharpoons H^+ + H_2PO_4^-; K_{a_1} = \frac{[H^+][H_2PO_4^-]}{[H_3PO_4]} = 1.1 \times 10^{-2} \qquad (130)$$

$$H_2PO_4^- \rightleftharpoons H^+ + HPO_4^{2-}; K_{a_2} = \frac{[H^+][HPO_4^{2-}]}{[H_2PO_4^-]} = 1.6 \times 10^{-7} \qquad (131)$$

$$HPO_4^{2-} \rightleftharpoons H^+ + PO_4^{3-}; K_{a_3} = \frac{[H^+][PO_4^{3-}]}{[HPO_4^{2-}]} = 4.8 \times 10^{-13} \qquad (132)$$

The pK_a values are pK_{a_1}, 1.96; pK_{a_2}, 6.8; and pK_{a_3}, 12.3, corresponding to the acid ionization constants, K_{a_1}, K_{a_2}, and K_{a_3}, respectively. Accordingly, the pH titration curve (Figure 4.2) shows a three-step profile. Buffer calculations are done separately for each conjugate pair present in the buffer. Exact calculations are difficult because at any pH within the buffering range of the acid, all the conjugate forms are present and require separate calculations. To simplify the calculations, approximations are introduced by ignoring those conjugate forms with concentrations that are too low relative to the others. As a guide, when the concentration is less than 1/100th of the total buffer concentration, that conjugate form is ignored in the calculations.

Examples: See problems 88c–d and 89b–c.

4.4. Practical Examples

PROBLEM 88

Prepare 0.800 L of 0.050 M solutions of the following buffers using calculated amounts of the conjugate acid and base. (a) Tris–acetate, pH 8.5 using Tris base (fw, 121.1) and 10.0 M acetic acid solution. (b) Acetate, pH 5.0 using sodium acetate trihydrate (fw, 136.08) and 10.0 M acetic acid solution. (c) Citrate (Cit), pH 5.0 using NaH_2Cit (fw, 214.1), Na_2HCit (fw, 236.1), and $Na_3Cit \cdot 2H_2O$ (fw, 294). (d) Phosphate, pH 7.2 using $NaH_2PO_4 \cdot H_2O$ (fw, 138) and $Na_2HPO_4 \cdot H_2O$ (fw, 268.07). The pK_a values are: Tris, 8.1; acetic acid, 4.8; citric acid, 3.1, 4.8, and 5.4 for pK_{a_1}, pK_{a_2}, and pK_{a_3}, respectively; phosphoric acid, 1.96, 6.8, and 12.3 for pK_{a_1}, pK_{a_2}, and pK_{a_3}, respectively.

SOLUTION:

(a) Preparation of Tris–acetate buffer using method 2.

At pH 8.5, Tris exists as the conjugate base, $Tris-NH_2$ and the conjugate acid, $Tris-NH_3^+$. Starting with Tris base which is 100% $Tris-NH_2$, $Tris-NH_3^+$ will be produced by adding acetic acid (HOAc) to convert $Tris-NH_2$ to $Tris-NH_3^+$ at a 1:1 equivalent ratio.

$$\therefore [HOAc] = [Tris-NH_3^+].$$

Calculate the following:

(i) the weight of Tris base needed; (ii) the $[Tris-NH_3^+]$ at pH 8.5; (iii) the milliliters of 10M HOAc needed. Using equation 9,

$$\text{g of Tris base needed} = M \times \text{mol wt} \times L$$
$$= 0.050 \text{ mol/L} \times 121.1 \text{ g/mol} \times 0.800 \text{ L}$$
$$= 4.80 \text{ g}$$

$[Tris-NH_3^+]$ is calculated using the Henderson-Hasselbach equation (equation 116):

$$pH = pK_a + \log \frac{[R-NH_2]}{[R-NH_3^+]}$$

When solved as shown in equation 119,

$$\frac{[R-NH_2]}{[R-NH_3^+]} = 10^{(pH-pK_a)}$$

Substituting 8.1 for pK_a and 8.5 for pH,

$$\frac{[Tris-NH_2]}{[Tris-NH_3^+]} = 10^{(8.5-8.1)} = 10^{0.4} = 2.51 \therefore [TrisNH_2] = 2.51 [Tris-NH_3^+] \quad (133)$$

Also,

$$[Tris-NH_2] + [Tris-NH_3^+] = 0.050 \text{ M (i.e., total } [Tris] = 0.050M)$$
$$\therefore [Tris-NH_2] = 0.050 - [Tris-NH_3^+] \quad (134)$$

Combining equations 133 and 134,

$$2.51 \, [\text{Tris-NH}_3^+] = 0.050 - [\text{Tris-NH}_3^+]$$

$$3.51 \, [\text{Tris-NH}_3^+] = 0.050 \, \text{M}$$

$$\therefore [\text{Tris} - \text{NH}_3^+] = \frac{0.050 \, \text{M}}{3.51} = 0.014 \, \text{M}$$

$$[\text{HOAc}] \text{ in the buffer} = [\text{Tris-NH}_3^+] = 0.014 \, \text{M}$$

V_1, the volume of the 10.0 M HOAc that will yield this concentration in the buffer, is calculated using equation 40:

$$V_1 = \frac{M_2 V_2}{M_1} = \frac{0.014 \, \text{M} \times 0.800 \, \text{L}}{10.0 \, \text{M}} = 0.00114 \, \text{L} = 1.1 \, \text{mL}$$

where V_2 is the final volume of the buffer; M_1 and M_2 are the concentrations of HOAc in the 10.0 M HOAc and the final buffer, respectively.

To prepare the buffer, dissolve 4.80 g of Tris base in 400.0 mL of H_2O, add 1.1 mL of 10.0 M HOAc, and adjust the volume to 800.0 mL with H_2O. The pH should be 8.5.

Note: HOAc (pK_a, 4.8) is completely ionized at pH 8.5 and does not contribute to the buffering capacity of the solution.

(b) Preparation of acetate buffer using method 2.

The conjugate base is acetate (OAc^-) supplied by NaOAc, and the conjugate acid is acetic acid (HOAc). [OAc^-] and [HOAc] at pH 5.2 are calculated and used to calculate the grams of NaOAc and the milliliters of 10.0 M HOAc needed. The pK_a of HOAc is 4.8. Using the Henderson-Hasselbach equation (equation 115),

$$\text{pH} = pK_a + \log \frac{[\text{OAc}^-]}{[\text{HOAc}]}$$

When solved as shown in equation 119,

$$\frac{[\text{OAc}^-]}{[\text{HOAc}]} = 10^{(\text{pH} - pK_a)}$$

Substituting 4.8 for pK_a and 5.0 for pH,

$$\frac{[\text{OAc}^-]}{[\text{HOAc}]} = 10^{(5.0-4.8)} = 10^{0.2} = 1.58 \therefore [\text{OAc}^-] = 1.58 \, [\text{HOAc}] \qquad (135)$$

Also,

$$[\text{OAc}^-] + [\text{HOAc}] = 0.050 \, \text{M} \text{ (i.e., total } [\text{OAc}^-] = 0.050 \text{M)}$$

$$\therefore [\text{OAc}^-] = 0.050 - [\text{HOAc}] \qquad (136)$$

Combining equations 135 and 136,

$$1.58 \, [\text{HOAc}] = 0.050 - [\text{HOAc}]$$

$$2.58 \, [\text{HOAc}] = 0.050 \text{ M}$$

$$\therefore [\text{HOAc}] = 0.050 \div 2.58 = 0.0194 \text{ M}$$

and

$$[\text{OAc}^-] = (0.050 - 0.0194) \text{ M} = 0.0306 \text{ M}$$

The grams of NaOAc·3H$_2$O needed are calculated using equation 9,

$$\text{g of NaOAc·3H}_2\text{O needed} = \text{M} \times \text{mol wt} \times \text{L}$$
$$= 0.0306 \text{ mol/L} \times 136.08 \text{ g/mol} \times 0.800 \text{ L}$$
$$= 3.30 \text{ g}$$

The volume of 10.0 M HOAc needed (V_1) is calculated using equation 39:

$$V_1 = \frac{M_2 V_2}{M_1} = \frac{0.0194 \text{ M} \times 0.800 \text{ L}}{10.0 \text{ M}} = 0.00155 \text{ L} = 1.6 \text{ mL}$$

where V_2 is the final volume of the buffer; M_1 and M_2 are the concentrations of HOAc in the 10.0 M HOAc and the final buffer, respectively.

To prepare the buffer, dissolve 3.30 g of NaOAc·3H$_2$O in 400.0 mL of H$_2$O, add 1.6 mL of 10.0 M HOAc, and adjust the volume to 800.0 mL with H$_2$O. The pH should be 5.0.

(c) Preparation of citrate buffer using method 2.

Citric acid (H$_3$Cit) is triprotic; pK_{a_1}, 3.1; pK_{a_2}, 4.8; pK_{a_3}, 5.4. At pH 5.0, most of the H$_3$Cit has dissociated to H$_2$Cit$^-$, HCit^{2-}, and Cit^{3-}. [H$_3$Cit] is too low (about 1/100 of the total [citrate]) and is ignored in the calculations. The dissociation equilibrium is, therefore,

$$\text{H}_2\text{Cit}^- \overset{K_2}{\rightleftharpoons} \text{HCit}^{2-} \overset{K_3}{\rightleftharpoons} \text{Cit}^{3-}$$

Calculate the molarity of each conjugate form present in the buffer, and the grams of NaH$_2$Cit, Na$_2$HCit, and Na$_3$Cit·2H$_2$O needed. Using the Henderson-Hasselbach equation (equation 101),

$$\text{pH} = \text{p}K_a + \log \frac{[\text{HCit}^{2-}]}{[\text{H}_2\text{Cit}^-]}$$

When solved as shown in equation 119,

$$\frac{[\text{HCit}^{2-}]}{[\text{H}_2\text{Cit}^-]} = 10^{(\text{pH}-\text{p}K_{a_2})}$$

Substituting 4.8 for pK_{a_2}, and 5.0 for pH,

$$\frac{[\text{HCit}^{2-}]}{[\text{H}_2\text{Cit}^-]} = 10^{(5.0-4.8)} = 10^{0.2} = 1.58$$

$$\therefore \ [H_2Cit^-] = \frac{[HCit^{2-}]}{1.58} = 0.633 \ [HCit^{2-}] \qquad (137)$$

A similar calculation for the second equilibrium yields

$$[Cit^{3-}] = 0.398 \ [HCit^{2-}] \qquad (138)$$

Also,

$$[H_2Cit^-] + [HCit^{2-}] + [Cit^{3-}] = 0.050 \ M \ (i.e., \ total \ [Cit] = 0.050 \ M) \qquad (139)$$

Substituting equations 137 and 138 into equation 139 yields:

$$0.633 \ [HCit^{2-}] + [HCit^{2-}] + 0.398 \ [HCit^{2-}] = 0.050 \ M$$

$$2.031 \ [HCit^{2-}] = 0.050 \ M$$

$$\therefore \ [HCit^{2-}] = 0.0246 \ M$$

Using equations 137 and 138, respectively,

$$[H_2Cit^-] = 0.0157 \ M,$$

and

$$[Cit^{3-}] = 0.00979 \ M$$

The grams of citrate salts needed are then calculated using equation 9:

$$g \ of \ NaH_2Cit = M \times mol \ wt \times L$$
$$= 0.0157 \ mol/L \times 214 \ g/mol \times 0.800 \ L = 2.68 \ g$$

Similarly,

$$g \ of \ Na_2HCit \ needed = 4.64 \ g, \ and$$

$$g \ of \ Na_3Cit \cdot 3H_2O \ needed = 2.30 \ g$$

To prepare the buffer, dissolve 2.68 g of NaH$_2$Cit, 4.64 g Na$_2$HCit, and 2.30 g of Na$_3$Cit·3H$_2$O in 400.0 mL of H$_2$O. Adjust volume to 800.0 mL. The pH should be 5.0.

(d) Preparation of phosphate buffer using method 2.

Phosphoric acid is triprotic with pK_{a_1}, 1.96; pK_{a_2}, 6.8; and pK_{a_3}, 12.3. At pH 7.4, only $H_2PO_4^-$ and HPO_4^{2-} are present in significant concentrations; when calculated, $[H_3PO_4]$ and $[PO_4^{3-}]$ are only about 10^{-5} times the total phosphate concentration. They are, therefore, ignored in the calculations. The predominant equilibrium is:

$$H_2PO_4^- \overset{K_2}{\rightleftharpoons} HPO_4^{2-}$$

The conjugate acid, $H_2PO_4^-$, and the conjugate base, HPO_4^{2-}, are supplied by NaH_2PO_4 and Na_2HPO_4, respectively. Calculate the $[H_2PO_4^-]$ and $[HPO_4^{2-}]$ at pH 7.4 and use the results to calculate the grams of NaH_2PO_4 and Na_2HPO_4 needed. Using the Henderson-Hasselbach equation (equation 115),

$$pH = pK_a + \log \frac{[HPO_4^{2-}]}{[H_2PO_4^-]}$$

When solved as shown in equation 119,

$$\frac{[HPO_4^{2-}]}{[H_2PO_4^-]} = 10^{(pH - pK_{a_2})}$$

Substituting 6.8 for pK_{a_2} and 7.4 for pH,

$$\frac{[HPO_4^{2-}]}{[H_2PO_4^-]} = 10^{(7.4-6.8)} = 10^{0.6} = 3.98 \therefore [HPO_4^{2-}] = 3.98[H_2PO_4^-] \qquad (140)$$

Also,

$$[HPO_4^{2-}] + [H_2PO_4^-] = 0.050 \text{ M (i.e., total [phosphate]} = 0.050 \text{ M)}$$

$$[HPO_4^{2-}] = 0.050 - [H_2PO_4^-] \qquad (141)$$

Combining equations 140 and 141 yields:

$$3.98\ [H_2PO_4^-] = 0.050 - [H_2PO_4^-]$$

$$4.98\ [H_2PO_4^-] = 0.050$$

$$[H_2PO_4^-] = \frac{0.050}{4.98} = 0.0100 \text{ M}$$

and

$$[HPO_4^{2-}] = (0.050 - 0.0100) = 0.040 \text{ M}$$

The grams of phosphate salts needed are calculated using equation 9:

$$g \text{ of } NaH_2PO_4 \cdot H_2O \text{ needed} = M \times \text{mol wt} \times L$$
$$= 0.0100 \text{ mol/L} \times 138 \text{ g/mol} \times 0.800 \text{ L} = 1.10 \text{ g}$$

Similarly,

$$g \text{ of } Na_2HPO_4 \cdot 7H_2O \text{ needed}$$
$$= 0.0400 \text{ mol/L} \times 268 \text{ g/mol} \times 0.800 \text{ L}$$
$$= 8.58 \text{ g}$$

Therefore, dissolve 1.10 g of NaH_2PO_4 and 8.58 g of Na_2HPO_4 in 400.0 mL of H_2O and adjust volume to 800.0 mL. The pH should be 7.4.

PROBLEM 89

Prepare 1.00 L of 0.100 M solutions of the following buffers by mixing appropriate volumes of the conjugate acid and base: (a) Acetate, pH 5.0, using sodium acetate trihydrate (fw 136.08) and concentrated acetic acid. (b) Phthalate, pH 6.0, using KH(phthalate) (fw 204.22) and K_2(phthalate) (fw 242.32). (c) Phosphate, pH 7.4, using $NaH_2PO_4 \cdot H_2O$ (fw 137.99) and $Na_2HPO_4 \cdot 7H_2O$ (fw 268.07). The pK_a values are: acetic acid, 4.8; phthalic acid, 2.95 and 5.41; phosphoric acid, 1.96, 6.8, and 12.3.

SOLUTION:

For each buffer, prepare 1.00 L of a 0.100 M solution of the conjugate base and acid separately. Calculate the [conjugate base]/[conjugate acid] ratio. Use this ratio to calculate the volumes of the conjugate acid and conjugate base needed to prepare 1.00 L of each buffer.

(a) Preparation of acetate buffer using method 3.

The conjugate base is acetate (OAc^-), and the conjugate acid is acetic acid (HOAc). Using equation 9,

g of $NaOAc \cdot 3H_2O$ required for 1.00 L of a 0.100 M solution = M × mol wt × L
$$= 0.100 \text{ mol/L} \times 136.08 \text{ g/mol} \times 1.00 \text{ L} = 13.61 \text{ g}$$

V_1, the volume of concentrated HOAc needed to prepare 1.00 L of the 0.100 M solution, is calculated using equation 39:

$$V_1 = \frac{M_2 V_2}{M_1} = \frac{0.100 \text{ N} \times 1.00 \text{ L}}{17.4 \text{ N}}$$

$$= 5.747 \times 10^{-3} \text{ L or 5.7 mL}$$

where V_2 is the final volume of the 0.100 M HOAc; M_1 and M_2 are the concentrations of HOAc in the concentrated and the 0.100 N acid solutions, respectively. **Prepare the solutions using these calculated amounts.** Calculate the [OAc^-]/[HOAc] ratio using equation 115:

$$pH = pK_a + \log \frac{[OAc^-]}{[HOAc]}$$

When solved as shown in equation 119, with $pK_a = 4.8$, and pH = 5.0,

$$\frac{[OAc^-]}{[HOAc]} = 10^{(pH-pK_a)} = 10^{(5.0-4.8)} = 10^{0.2} = 1.58$$

This means that 1.58 parts of OAc^- is present for every part of HOAc, a total of 2.58 parts. Thus,

1 part of 1.00 L buffer = (1000.0 mL)/2.58 = 387.6 mL

∴ vol of 0.100 M HOAc needed = 388.0 mL

and

vol of 0.100 M NaOAc⁻ needed = 387.6 mL × 1.58 = 612.0 mL

∴ **Mix 388.0 mL of the 0.100 M HOAc and 612.0 mL of the 0.100 M NaOAc to yield 1.00 L of the 0.100 M acetate buffer, pH 5.0.**
(b) Preparation of phthalic buffer using method 3.
First, calculate the grams of KH(phthalate) and K_2(phthalate) needed to prepare 1.00 L of a 0.100 M solution of each (equation 9):

$$\text{g of KH(phthalate) needed} = \text{M} \times \text{mol wt} \times \text{L}$$
$$= 0.100 \text{ mol/L} \times 204.22 \text{ g/mol} \times 1.00 \text{ L}$$
$$= 20.40 \text{ g}$$

Similarly,

$$\text{g of } K_2\text{(phthalate) needed}$$
$$= 0.100 \text{ mol/L} \times 242.32 \text{ g/mol} \times 1.00 \text{ L}$$
$$= 24.21 \text{ g}$$

Prepare the solutions using these calculated amounts. Next, calculate the $[\text{pht}^{2-}]/[\text{Hpht}^-]$ ratio: Phthalic acid (H_2pht) is diprotic; pK_{a_1}, 2.95; pK_{a_2}, 5.41. It dissociates in two steps:

$$H_2\text{pht} \overset{K_{a_1}}{\rightleftharpoons} \text{Hpht}^- \overset{K_{a_2}}{\rightleftharpoons} \text{pht}^{2-}$$

At pH 6.0, most of the H_2pht has dissociated to Hpht⁻. The remaining $[H_2\text{pht}]$ is too low and is ignored in the calculations. The $[\text{pht}^{2-}]/[\text{Hpht}^-]$ ratio is calculated using the Henderson-Hasselbach equation (equation 115):

$$\text{pH} = pK_a + \log \frac{[\text{pht}^{2-}]}{[\text{Hpht}^-]}$$

When solved as shown in equation 119, with $pK_{a_2} = 5.41$, pH = 6.0,

$$\frac{[\text{pht}^{2-}]}{[\text{Hpht}^-]} = 10^{(\text{pH-}pK_{a_2})} = 10^{0.59} = 3.89$$

This means that 3.89 parts of pht^{2-} is present for every part of Hpht⁻, a total of 4.89 parts.

$$1 \text{ part of } 1.00 \text{ L buffer} = (1000.0 \text{ mL})/4.89 = 204.5 \text{ mL}$$

$$\therefore \text{ vol of } 0.100 \text{ M KHpht needed} = 204.0 \text{ mL}$$

and

$$\text{vol of } 0.100 \text{ M } K_2\text{pht needed} = 204.5 \text{ mL} \times 3.89 = 796.0 \text{ mL}$$

∴ **Mix 204.0 mL of the 0.100 M KHpht and 796.0 mL of the 0.100 M K_2pht to yield 1.00 L of the 0.100 M phthalate buffer, pH 6.0.**

(c) Preparation of phosphate buffer using method 3.

First, use equation 9 to calculate the grams of NaH_2PO_4 and Na_2HPO_4 needed to prepare 1.00 L of a 0.100 M solution of each:

$$\text{g of } NaH_2PO_4 \cdot H_2O \text{ needed} = M \times \text{mol wt} \times L$$
$$= 0.100 \text{ mol/L} \times 138 \text{ g/mol} \times 1.00 \text{ L}$$
$$= 13.80 \text{ g}$$

Similarly,

$$\text{g of } Na_2HPO_4 \cdot 7H_2O \text{ needed}$$
$$= 0.100 \text{ mol/L} \times 268 \text{ g/mol} \times 1.00 \text{ L}$$
$$= 26.80 \text{ g}$$

Prepare the 0.100 M solutions using these calculated amounts.

Next, calculate the $[HPO_4^{2-}]/[H_2PO_4^{-}]$ ratio:

H_3PO_4 dissociates in three steps, and at pH 7.4, the predominant conjugate acid is $H_2PO_4^{2-}$, and the predominant conjugate base is HPO_4^{2-} (see problem 88d for details). Calculate the $[HPO_4^{2-}]/[H_2PO_4^{2-}]$ ratio using the Henderson-Hasselbach equation (equation 115),

$$pH = pK_a + \log \frac{[HPO_4^{2-}]}{[H_2PO_4^{2-}]}$$

When solved as shown in equation 119, with $pK_{a_2} = 6.8$, and pH 7.4,

$$\frac{[HPO_4^{2-}]}{[H_2PO_4^{2-}]} = 10^{(pH-pK_{a_2})} = 10^{0.6} = 3.98$$

This means that 3.98 parts of HPO_4^{2-} is present for every part of $H_2PO_4^{-}$, a total of 4.98 parts. Thus,

$$1 \text{ part of } 1.00 \text{ L buffer} = (1000.0 \text{ mL})/4.98 = 200.8 \text{ mL}$$

$$\therefore \text{ vol of } 0.100 \text{ M } NaH_2PO_4 \text{ needed} = 200.8 \text{ mL}$$

and

$$\text{vol of } 0.100 \text{ M } Na_2HPO_4 \text{ needed} = 200.8 \text{ mL} \times 3.98 = 799.0 \text{ mL}$$

To prepare the buffer, mix 200.8 mL of the 0.100 M NaH_2PO_4 and 799.0 mL of the 0.100 M Na_2HPO_4 to yield 1.00 L of the 0.100 M phosphate buffer, pH 7.4.

PROBLEM 90

The pH of a 2.0 mL assay medium buffered with 0.050 M glycylglycine (pK_a 8.4) decreased from 8.5 to 8.3 at the end of the assay. Calculate the moles of H^+ released into the medium.

SOLUTION:

Calculate the change in the total $[H^+]$ in the buffer due to the change in pH from 8.5 to 8.3; i.e. $(d[H^+]_T)$. From equation 126,

$$d[H^+]_T = \frac{C}{1+10^{(pH_2-pK_a)}} - \frac{C}{1+10^{(pH_1-pK_a)}} + (10^{-pH_2} - 10^{-pH_1})$$

$$= \frac{0.050 \text{ M}}{1+10^{(8.3-8.4)}} - \frac{0.050 \text{ M}}{1+10^{(8.5-8.4)}} + (10^{-8.3} - 10^{-8.5})$$

where C, the total buffer concentration = 0.050 M, pK_a = 8.4, pH_1 = 8.5, and pH_2 = 8.3.

$$\therefore d[H^+]_T = (0.0279 - 0.0221) \text{ M} + (1.85 \times 10^{-9} \text{ M}) = 0.0058 \text{ M}$$
$$= (0.0279 - 0.0221) \text{ M} = 0.0058 \text{ M}$$

H^+ released into the 2.0 mL assay medium =

$$\frac{0.0058 \text{ mol}}{1000.0 \text{ mL}} \times 2.0 \text{ mL} = 11.6 \times 10^{-6} \text{ mol or 12 } \mu\text{mol}$$

PROBLEM 91

What is the pH of (a) 0.0050 M HCl, and (b) 0.100 M HCl.

SOLUTION:

In dilute solutions, strong acids and bases ionize completely. Therefore,

$$[H^+] = [HCl].$$

(a) $[H^+]$ = 0.0050 M. Using equation 97,

$$pH = -\log [H^+] = -\log 0.0050 = 2.3$$

(b) $[H^+]$ = 0.100

$$pH = -\log 0.100 = 1.0$$

Note: The pH of the solution depends on the concentration of the acid, and unlike weak acids (or weak bases), strong acids (or strong bases) do not possess buffering properties.

PROBLEM 92

A 0.200 L aliquot of 0.050 M solution of dibasic sodium phosphate (Na_2HPO_4) is mixed with 0.160 L of 0.050 M monobasic sodium phosphate (NaH_2PO_4). Calculate the pH of the resulting buffer.

SOLUTION:

The conjugate acid is NaH_2PO_4, and the conjugate base is Na_2HPO_4. Since the concentrations of the two solutions are equal,

$$\frac{[NaHPO_4]}{[NaH_2PO_4]} = \frac{0.200 \text{ L}}{0.160 \text{ L}} = 1.250$$

Using equation 115,

$$pH = pK_a + \log \frac{[NaHPO_4]}{[NaH_2PO_4]} = 6.8 + \log 1.250 = 6.9$$

PROBLEM 93

Carbonic acid, H_2CO_3, is a weak diprotic acid formed by the reaction of carbon dioxide with water. It has $K_{a_1} = 4.3 \times 10^{-7}$ and $K_{a_2} = 5.6 \times 10^{-11}$.

What are the equilibrium concentrations in a 0.10 M solution of the acid?

SOLUTION:

From the two values of the K_a, K_{a_1} is much larger than K_{a_2}, and we can assume that nearly all the hydrogen ion in the solution is derived from the first ionization.

$$H_2CO_3 \rightleftharpoons H^+ + HCO_3^- \qquad\qquad K_{a_1} = 4.3 \times 10^{-7}$$

$$HCO_3^- \rightleftharpoons H^+ + CO_3^{-2} \qquad\qquad K_{a_2} = 5.6 \times 10^{-11}$$

As also seen from the value of K_{a_2}, only very little of HCO_3^- will undergo further dissociation.

For the first dissociation

$$K_{a_1} = \frac{[H^+][HCO_3^-]}{[H_2CO_3]}$$

Let x equals the number of moles per liter of H_2CO_3 that dissociates. From the stoichiometry of the first step, we obtain x mol/L of H^+ and x mol/L of HCO_3^-. At equilibrium, there will be $(0.10 - x)$ mol/L of H_2CO_3 remaining.

Buffer components	Initial molar concentrations	Change	Equilibrium molar concentrations
H^+	0.0	+x	x
HCO_3^-	0.0	+x	x
H_2CO_3	0.10	−x	0.10 − x = 0.10

As before since K_{a_1} is small we can assume x will be negligible compared to 0.10, thus

$$[H_2CO_3] = .10 - x \approx 0.10 \text{ M}$$

Substituting these equilibrium quantities into the expression for K_{a_1}, we have

$$\frac{(x)(x)}{0.10} = K_{a_1} = 4.3 \times 10^{-7}$$

$$x^2 = 4.3 \times 10^{-8}$$

$$x = 2.1 \times 10^{-4}$$

∴ The equilibrium concentrations from the first dissociation are

$$[H^+] = 2.1 \times 10^{-4} \, M$$

$$[HCO_3^-] = 2.1 \times 10^{-4} \, M$$

$$[H_2CO_3] = 0.10 - 2.1 \times 10^{-4} = 0.10 \, M$$

Using K_{a_2}, we can calculate the equilibrium concentration of CO_3^{-2}

$$HCO_3^- \rightleftharpoons H^+ + CO_3^{-2}$$

$$K_{a_2} = \frac{[H^+][CO_3^{-2}]}{[HCO_3^-]}$$

Let y equal the number of moles per liter of HCO_3^- that dissociates. Then, from the stoichiometry of this second dissociation step, additional moles per liter of H^+ and CO_3^{-2} produced will also be y. The total hydrogen ion concentration from both first and second dissociations will be $[H^+] = (2.1 \times 10^{-4} + y)$, and the concentration of HCO_3^- that remains at equilibrium will be $(2.1 \times 10^{-4} - y)$. Thus, for the second dissociation,

Buffer components	Initial molar concentrations	Change	Equilibrium molar concentrations
H^+	2.1×10^{-4}	$+y$	$2.1 \times 10^{-4} + y$
HCO_3^{-2}	0.0	$+y$	y
HCO_3^-	2.1×10^{-4}	$-y$	$2.1 \times 10^{-4} - y$

Since K_{a_2} is really very small, the amount of HCO_3^- that will dissociate will also be small and the assumption that y is negligible compared to 2.1×10^{-4} is made, to give the equilibrium concentration as

$$[H^+] = 2.1 \times 10^{-4} + y \approx 2.1 \times 10^{-4} \, M$$

$$[CO_3^{-2}] = y$$

$$[HCO_3^-] = 2.1 \times 10^{-4} - y \approx 2.1 \times 10^{-4} \, M$$

which on substitution gives

$$K_{a_2} = \frac{(2.1 \times 10^{-4})(y)}{(2.1 \times 10^{-4})} = 5.6 \times 10^{-11}$$

$$y = 5.6 \times 10^{-11}$$

$$\therefore [CO_3^{-2}] = 5.6 \times 10^{-11} \, M$$

We can summarize the concentrations of the solute species present at equilibrium in 0.10 M H_2CO_3 as

$$[H^+] = 2.1 \times 10^{-4} \text{ M}$$

$$[HCO_3^-] = 2.1 \times 10^{-4} \text{ M}$$

$$[CO_3^{-2}] = 5.6 \times 10^{-11} \text{ M}$$

$$H_2CO_3 = 0.10 \text{ M}$$

Chapter 5

Spectrophotometry: Basic Principles and Quantitative Applications

5.1. Principles and Techniques

Spectrophotometry is the science of measuring the light absorption characteristics of substances for use in determining their concentration, identity, and other properties. This technique is possible because many substances absorb light of specific wavelengths within the ultraviolet (200–400 nm), visible (400–700 nm), and near-infrared (700–1000 nm) regions of the electromagnetic spectrum, and transmit the remainder. The wavelength of the transmitted (or reflected) light is what imparts a characteristic color to a substance: For example, a red wine appears red because it transmits red light (wavelength, 600–700 nm), and absorbs light of shorter wavelengths.

A measure of the amount of light absorbed by a substance is called **absorbance** or **optical density (OD)**, and a measure of the amount transmitted is called **transmittance** (see Appendix A.3). The wavelength at which a substance has maximum absorbance is characteristic of the substance and is used to identify the substance. If two substances contain the same light-absorbing prosthetic group (e.g., riboflavin and flavin adenine dinucleotide both contain flavin), they usually absorb maximally at about the same wavelength, but the absorbance per mole of each substance **(molar extinction coefficient)** is usually different. A recording of the absorbance of a substance as a function of wavelength is called the **absorption spectrum** (Figure 5.1). The peaks (**absorption maxima**) on an absorbance spectrum are the regions of high absorption, and the troughs (**absorption minima**) are the regions of low absorption.

Absorbance is measured with a spectrophotometer, and the basics of this technique and the instrumentation are described in Appendix A.4. The rest of this chapter deals with basic calculations and applications of this versatile technique.

5.2. Quantitative Aspects and Basic Applications

A. The Beer-Lambert Law

This law is the basis for applying spectrophotometry to the calculation of concentrations. It states that the absorbance (A) of a substance in solution is linearly proportional to its concentration (C) and the length (ℓ) of the light path through the solution. The law is expressed mathematically (see Appendix A.3 for derivation) as follows:

$$A = E\ell C \qquad (142)$$

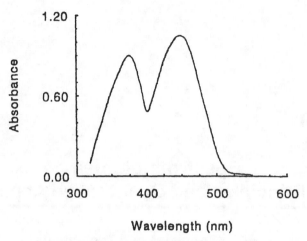

Figure 5.1.   A spectrum of riboflavin in the visible and near-UV regions.

Rearranging equation 142,

$$C = \frac{A}{E\ell} \tag{143}$$

and

$$E = \frac{A}{\ell C} \tag{144}$$

The proportionality constant, E, is characteristic of the substance at the wavelength in which the absorbance was measured. It is known by various other names: **extinction coefficient**, **absorption coefficient**, **absorptivity**, and **absorbency index**. Its unit and symbolic designation depend on the units of C and ℓ. When C and ℓ are 1 M and 1 cm, respectively, E is designated E_{1cm}^{1M} (λ) or ϵ, the molar extinction coefficient; otherwise, the general designation used is $E\ell C$ (λ). Equation 143 is used to directly calculate the concentration of a substance once A, E, and ℓ are known. Equation 144 is used to calculate E for a substance at a given wavelength. To do this, the A of a known concentration of the substance is measured at this wavelength, using a cuvette with a known ℓ. The A, C, and ℓ are then used to calculate the E. The A can, alternatively, be obtained from a spectrum of a known concentration of the substance.

B. Calculating Concentrations

(1) Nonenzymatic analytes

The Beer-Lambert law is used to calculate the concentration of substances using absorbance data and equation 143. The calculation is handled in two ways:

(a) Direct calculation

This is applicable to substances that directly absorb light. To calculate the concentration of a sample, first measure its absorbance. Obtain E and ℓ; if unknown, measure E as described above, and measure ℓ as the length of the cuvette's light path (usually 1 cm). Use equation 143 to calculate the concentration.

Examples: See problems 94, 95a, 97a, and 98.

(b) Use of a standard curve to calculate concentration

The use of a standard curve is often necessary when analyzing substances that do not absorb light directly. These substances may be converted to light-absorbing species by either a chemical reaction or by coupling to other reactions in which a light absorber is being consumed or produced. Direct calculation of concentration in these cases is tedious and sometimes impossible because the extinction coefficients of the light-absorbing species are unknown. Standards of the pure substance are assayed along with the samples, and a standard curve is constructed by plotting the absorbance values for these standards against the corresponding concentrations. The concentration of the sample is then obtained by reading from this curve the concentration corresponding to the sample's absorbance. If the sample was diluted, the result is multiplied by the dilution factor. However, if the standards and the sample were diluted equally, the concentration of the sample should not be multiplied by the dilution factor since this factor was incorporated into the standard curve.

Example: See problem 99.

Alternatively, the absorbance and concentration of a single standard can be used to calculate the unknown concentration, provided that this standard is within the linear range of the assay.

Example: See problem 99.

A standard curve can also be used when several samples are being assayed simultaneously; reading the concentrations off the standard curve is faster than calculating them.

(2) Analyzing enzymes spectrophotometrically

The concentration of an enzyme is usually measured as activity. Because activity is defined as the amount of substrate consumed or product liberated per unit time, the activity of an enzyme can be determined spectrophotometrically if the substrate or product absorbs light, or if it can be coupled to a reaction in which a light absorber is being consumed or produced. Enzymes that catalyze oxidation-reduction (redox) reactions (e.g., the cytochromes), contain light-absorbing prosthetic groups. Their redox transition activities are best assayed spectrophotometrically (*4,5*).

Examples: See problems 97, 98, 100, 102, and Chapter 6.

Notes: (a) The Beer-Lambert equation is linear and has zero intercept and a slope of $E\ell$; therefore, the standard curve should pass through the zero origin. (b) At higher concentrations, the curve usually becomes nonlinear mainly because of intermolecular interactions, depletion of light, light scattering, instrument limitations at low levels of transmitted light, and reagent depletion. Absorbance values above the linear limit do not give accurate concentrations. Therefore, **(i)** only the linear portion of a standard curve should be used to obtain the concentration of samples; samples with higher absorbance values should be diluted and re-assayed. **(ii)** Except when a sophisticated spectrophotometer with high accuracy is used, absorbance values greater than about 1.5 should not be used for calculations; the sample should be diluted and re-assayed.

C. Quantifying Nucleic Acids

Nucleic acids absorb UV light between 250 and 280 nm, with DNA and RNA absorbing optimally at 260 nm (see Appendix F for specifics). As such, they can be quantified spectrophotometrically using equation 142. However, molecular biologists usually quantify nucleic acids in **absorbance units (AU)** or **optical density (OD)** units. One AU or OD unit of a substance (e.g., nucleic acid) is defined as the concentration that gives 1 AU at a given wavelength. The quantitative basis for this definition is obtained by rearranging equation 142:

$$OD = E\ell C$$

By rearranging and substituting 1 cm for ℓ :

$$\frac{C}{OD} = \frac{1}{E} = 1 \text{ AU} \tag{145}$$

C/OD defines 1 AU or 1 OD unit. Thus, **1 AU is the inverse of the absorption coefficient.** Equation 145 is used to obtain the value of 1 AU when the absorption coefficient is known, and vice versa. To calculate the concentration of nucleic acids (or other light-absorbing substances) using their AU and measured OD, the following equations can be used:

From equation 145,

$$\frac{C}{OD} = 1 \text{ AU}$$

$$\therefore C = 1 \text{ AU} \times OD$$

If the sample was diluted before measuring the OD and DF is the dilution factor, then

$$C = 1 \text{ AU} \times OD \times DF \tag{146}$$

Note: The unit for concentration will be that of the AU, usually mg/mL.

Values for the AU and absorption coefficient for RNA and DNA are given in Table 5.1, below. Additional data are given in Appendix F. To use these values to calculate concentrations, apply equation 146 when AU is known or equation 142 when the absorption coefficient is known.

Examples: See problems 95, 96, and 108.

Base pairing and stacking decrease the absorbance of DNA and RNA. Thus, when double-stranded DNA melts, its absorbance increases [**hyperchromic effect**, (6)]. For this reason, double-stranded DNA (or RNA) has a lower absorption coefficient and a higher value for AU than the single-stranded forms.

The values for absorption coefficient and AU for short single-stranded oligonucleotides vary, depending on the length and base composition: The shorter the oligo, the higher the absorption coefficient.

Table 5.1. Absorption Constants for Selected Nucleic Acids[a]

Nucleic acid	$E_{1cm}^{1mg/mL}$	1AU$_{260}$ (260 nm)(μg/mL)
Double-stranded DNA[b]	20	50
Single-stranded RNA	25	40
Single-stranded DNA[c]	$\approx$25	$\approx$40
Single-stranded oligos[d]	$\approx$30	$\approx$33

[a] See Appendix F for more data.
[b] Greater than 100 nucleotides.
[c] $\geq$50 nucleotides.
[d] $\leq$50 nucleotides.

When quantifying nucleic acids, take OD readings at 260 nm and 280 nm wavelengths. Calculate the concentration using the OD_{260}, then calculate the OD_{260}/OD_{280} ratio and use it to assess the purity of the sample: Pure DNA and RNA have OD_{260}/OD_{280} ratio of 1.8 and 2.0, respectively. If the ratios are significantly less than these values, then the samples are contaminated.

D. Identifying Substances

All direct absorbers have characteristic absorption spectra (7). The spectrum may be regarded as an optical "fingerprint" which identifies the compound. Identification is made using important features of the spectrum: **(i)** The wavelengths at which the absorption peaks and troughs occur are characteristic of the substance, **(ii)** the extinction coefficients at the absorption maxima are also characteristic, and **(iii)** the ratio of the absorbance at two characteristic wavelengths is a constant. As an example, the spectrum of riboflavin (Figure 5.1) shows absorption maxima at 450 and 375 nm, and an absorption minima at 400 nm. The E_{1cm}^{1M} (450 and 374 nm) are always 1.22×10^4 and 1.06×10^4, respectively, and the A_{450}/A_{375} ratio is always 1.15 for pure riboflavin. Together, these characteristics distinguish riboflavin from other compounds, including flavin adenine dinucleotide (FAD) which also contains flavin, the light-absorbing prosthetic group in riboflavin. Other compounds can be identified by converting them to light-absorbing analogues, and then recording and analyzing their spectra.

E. Characterizing Substances

Spectrophotometric techniques have been used to characterize biomolecules. Examples include oxidation-reduction transitions of heme proteins (4), the transitions of DNA between single- and double-stranded forms (6), ligand-receptor interactions including enzyme substrate interactions (7), and the determination of the pK_a of weak acids (8).

5.3. Practical Examples

PROBLEM 94

The absorbance of a solution of riboflavin in a 1.0 cm cuvette was 0.610 at 450 nm and 0.530 at 375 nm. Calculate the concentration of riboflavin in the solution. The E_{1cm}^{1M} of riboflavin at 450 nm and 375 nm are 1.22×10^4 and 1.06×10^4, respectively.

SOLUTION:

The concentration is calculated using either the A and E at 450 nm or the A and E at 375 nm. From equation 143,

$$C = \frac{A}{E\ell}$$

Using the data at 450 nm,

$$C = [\text{riboflavin}] = \frac{0.610}{1.22 \times 10^4 \text{ M}^{-1}\text{cm}^{-1} \times 1.0 \text{ cm}} = 5.0 \times 10^{-5} \text{ M}$$

Using the data at 375 nm

$$C = [\text{riboflavin}] = \frac{0.530}{1.06 \times 10^4 \text{ M}^{-1}\text{cm}^{-1} \times 1.0 \text{ cm}} = 5.0 \times 10^{-5} \text{ M}$$

PROBLEM 95

(a) Estimate the amount of DNA in a 20.0 mL sample if the absorbance at 260 nm of 10.0 µL of the sample in 1.99 mL of H_2O is 0.500. Use an $E_{1cm}^{1mg/mL}$ (260 nm) of 20.3 for DNA; light path, 1.0 cm. (b) An aliquot of a 2.0 mL solution of a purified protein was diluted 10-fold before 0.10 mL was mixed with 5.0 mL of Bradford reagent. If this gives an absorbance of 0.45 at 595 nm, estimate the milligrams of protein in the 2.0 mL solution given that 0.10 mL of a standard (1.0 mg/mL), similarly treated, gave a 0.510 absorbance.

SOLUTION:

(a) From equation 143,

$$C = \text{mg DNA/mL} = A/E\ell$$

$$\text{mg DNA / mL} = \frac{0.500}{20.3 \text{ (mg / mL)}^{-1}\text{cm}^{-1} \times 1.0 \text{ cm}} = 0.0246 \text{ mg/mL}$$

$$= \text{concentration of DNA in the assay medium}$$

To calculate mg DNA/mL of sample, let V_1 be the volume of the sample (10.0 µL) added to the H_2O, and C_1 its DNA concentration; V_2 and C_2 are the final volume of the assay medium and the concentration of DNA in it, respectively. Using equation 39,

$$C_1 = \frac{C_2 V_2}{V_1} = \frac{0.0246 \text{ mg / mL} \times 2.0 \text{ mL}}{0.0100 \text{ mL}} = 4.92 \text{ mg / mL}$$

This means that 1.0 mL of the sample contains 4.92 mg of DNA.

$$\therefore \ 20.0 \text{ mL of sample contains } \frac{4.92 \text{ mg}}{1 \text{ mL}} \times 20.0 \text{ mL} = 98.00 \text{ mg}$$

(b) Since both the sample and the standard were treated similarly, the milligrams of protein in the assay medium can be calculated by proportion:

$$\frac{\text{absorbance of standard}}{\text{mg / mL of standard}} = \frac{\text{absorbance of sample}}{\text{mg / mL of sample}}$$

$$\text{mg / mL of sample} = \frac{\text{absorbance of sample} \times \text{mg / mL of standard}}{\text{absorbance of standard}} \times 10$$

$$\text{mg of sample} = \frac{0.45 \times 1.0 \text{ mg / mL}}{0.510} \times 10 = 8.82 \text{ mg / mL}$$

$\therefore$ mg of protein in the 2.0 mL sample = 8.82 mg/mL $\times$ 2.0 mL = 18.00 mg

PROBLEM 96

You received a vial containing a 0.2 mL solution of a 20-base single-stranded DNA. If the concentration of DNA is given as 0.450 OD_{260} units, convert this concentration to (a) μg/mL and (b) pM. (c) Calculate the total amount of DNA in pmol. For a 20-base single-stranded DNA, assume that 1 $AU_{260} \approx 33$ μg DNA/mL. (See also problem 107.)

SOLUTION:

(a) Using equation 146,

$$\text{concn of DNA} = 1 \text{ AU} \times \text{OD} \times \text{DF}$$

$$33 \text{ μg/mL AU} \times 0.450 \text{ AU} \times 1 = 14.85 \text{ μg DNA/mL}$$

(b) Using information from Appendix F, the molecular mass of the DNA $\approx$ 325 D/base $\times$ 20 bases = 6500 D.

Using equation 10,

$$M = \frac{wt}{\text{mol wt} \times L} = \frac{1.485 \times 10^{-5} \text{ g}}{6500 \text{ g / mol} \times 2.0 \times 10^{-4} \text{ L}} = 1.142 \times 10^{-5} \text{ mol / L}$$

$$= 1.152 \times 10^{-5} \text{ mol / L} \times \frac{1 \text{ μM}}{1 \times 10^{-6} \text{ mol / L}} = 11.4 \text{ μM}$$

(c) Using equation 9,

$$M = \text{mol/L}$$

$$\therefore \text{ Total moles of DNA} = M \times L$$
$$= 1.142 \times 10^{-5} \text{ mol/L} \times 2.0 \times 10^{-4} \text{ L}$$
$$= 2.284 \times 10^{-9} \text{ mol}$$

$$= 2.284 \times 10^{-9} \text{ mol} \times \frac{1 \text{ pmol}}{1 \times 10^{-12} \text{ mol}} = 2.28 \times 10^{3} \text{ pmol}$$

PROBLEM 97

Cytochrome o oxidizes ubiquinol-1 (Q_1H_2) to ubiquinone-1 (Q_1) which absorbs light at 262 nm. After adding 80.0 μg of cytochrome o to an assay medium containing a saturating concentration of Q_1H_2, the absorbance changed from 0.010 to 0.190 in 2.0 min. Calculate

(a) moles of Q_1H_2 oxidized and (b) the specific activity of cytochrome o. Total volume of the assay medium was 2.0 mL, E_{1cm}^{1mM} (262 nm) of Q_1 is 15.0, and ℓ is 1.0 cm. Oxidized is abbreviated ox.

SOLUTION:

(a) The moles of Q_1H_2 oxidized equal the moles of Q_1 formed, and is represented by the change in absorbance (ΔA). Using equation 143,

$$C = [Q_1H_2]_{ox} = \frac{\Delta A}{E\ell} = \frac{(0.190 - 0.010)}{15.0 \ (mmol \, / \, L)^{-1} cm^{-1} \times 1.0 \ cm}$$

The total amount of oxidized Q_1H_2 in the 2.0mL assay medium

$$= \frac{(0.190 - 0.010)}{15.0 \ mM^{-1}cm^{-1} \times 1.0 \ cm} = \frac{2.0 \ mL}{1000.0 \ mL \, / \, L}$$

$$= 2.4 \times 10^{-5} \ mmol \ or \ 24 \ nmol$$

(b) Using equation 158,

$$sp \ act. \ of \ cytochrome \ o = \frac{amount \ of \ Q_1H_2 \ oxidized}{min \times mg \ of \ protein}$$

$$= \frac{24 \ mol}{2.0 \ mol \times 0.0800 \ mg \ protein}$$

$$= 150 \ nmol \ min^{-1} mg^{-1} \ protein$$

PROBLEM 98

A 0.1 mL aliquot of bacterial membrane preparation decreased the 340 nm absorbance of NADH from 0.415 to 0.138 in 5.0 min, in a 2.0-mL assay medium. Calculate the specific and total activities of NADH dehydrogenase in 100.0 mL of the extract containing 2.00 mg of protein/mL. Assume a saturating concentration of NADH, and a 1.0 cm light path. The E_{1cm}^{1M} (340 nm) of NADH is 6220.

SOLUTION:

The change in absorbance (ΔA) represents the total amount of NADH oxidized. By using equation 143,

$$C = [NADH]_{ox} = \frac{\Delta A}{E\ell} = \frac{(0.415 - 0.138)}{6220 \ M^{-1}cm^{-1} \times 1.0 \ cm}$$

The total amount of oxidized NADH in the 2.0 mL assay medium

$$= \frac{(0.415 - 0.138)}{6220 \ M^{-1}cm^{-1} \times 1.0 \ cm} = \frac{2.0 \ mL}{1000.0 \ mL \, / \, L}$$

$$= 8.91 \times 10^{-8} \ mol \ or \ 0.0890 \ \mu mol$$

Total protein in the assay medium

$$= 2.0 \text{ mg/mL} \times 0.10 \text{ mL} = 0.20 \text{ mg}$$

Using equation 158,

$$\text{sp act.} = \frac{0.0891 \ \mu\text{mol of NADH}}{5.0 \ \text{min} \times 0.20 \ \text{mg of protein}}$$

$$= 0.089 \ (\mu\text{mol min}^{-1}) \ \text{mg}^{-1} \text{ of protein}$$

Total protein/100.0 mL extract $= 2.0 \text{ mg/mL} \times 100.0 \text{ mL} = 200.00 \text{ mg}$

$\therefore$ Total activity in 100.0 mL extract

$$= 0.0891 \ (\mu\text{mol min}^{-1}) \ \text{mg}^{-1} \text{ protein} \times 200 \text{ mg of protein}$$

$$= 18 \ \mu\text{mol/min or 18 IU.}$$

where 1 International Unit (IU) = 1 μmol/min (see equation 159).

PROBLEM 99

p-Nitrophenol (PNP) absorbs light at 405 nm. A 0.1 mL aliquot of a sample containing PNP was added to 1.9 mL of H_2O and the absorbance measured at 405 nm was 0.550. The absorbance for standards treated similarly are as shown below. (a) What is the concentration of PNP in the sample? (b) If only the 30.0 μM standard is assayed, calculate the concentration of PNP in the sample. Assume that the extinction coefficient of PNP is unknown.

μM of standard	0.0	10.0	20.0	30.0	40.0
OD	0.000	0.180	0.421	0.592	0.810

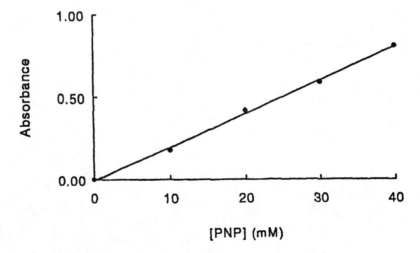

Figure 5.2. OD versus concentration standard curve for PNP.

SOLUTION:

(a) Plot the μM PNP versus OD (Figure 5.2). The concentration of PNP in the sample is the μM PNP on the standard curve corresponding to 0.550 AU. This value is 27.5 μM.

(b) Since the 30.0 μM standard is within the linear range of the assay (see standard curve), it can be used to calculate the [PNP] in the sample by simple proportion:

$$\frac{A_{sample}}{C_{sample}} = \frac{A_{standard}}{C_{standard}}$$

$$C_{sample} = \frac{A_{sample} \times C_{standard}}{A_{standard}} = \frac{0.550 \times 30.0 \ \mu M}{0.592} = 27.0 \ \mu M$$

Chapter 6

Enzyme Assays and Activity

6.1. Introduction

Every enzyme assay involves the use of an enzyme as a catalyst to convert a substrate (and co-substrate in some cases) to product. To detect the reaction, a property (e.g., light absorption) of either the substrate, product, or co-substrate is measured at appropriate times. If none of them has a suitable property, the reaction can be coupled to a second reaction system that has a readily measurable property. Coupling is achieved by making a product of the first reaction serve as a substrate or co-substrate for the second. The stoichiometries of the reactions facilitate calculating the amount of substrate or product catalyzed in the first reaction.

Enzyme assays may be qualitative or quantitative. The qualitative assay seeks to establish the presence or absence of either the enzyme or the substrate in a sample. In contrast, the quantitative assay measures the amount of enzyme or substrate in a sample and requires optimal conditions if accuracy is desired. Quantitative assays can be classified into two broad categories: One category includes those that are aimed at determining the enzyme's activity or kinetic constants, and the other category includes those that are aimed at determining the concentration of substrate, using the enzyme as a reagent.

Applications for enzyme assays include the determination of the activity and kinetic constants for enzymes; determination of substrate or co-substrate concentration; determination of antigen concentration by means of enzyme immunoassays; qualitative detection of biomolecules, cellular components, cells, tissues, or other matrices in which a suitable enzyme or substrate serves as a marker.

The early parts of this chapter outline the quantitative basis for establishing conditions for enzyme assays. The later parts discuss how the raw data obtained from these assays are used to calculate activities and other parameters for enzymes. Definitions of activities, activity units, and kinetic constants are summarized.

6.2. Conditions Required for Quantitative Enzyme Assays

A. The Michaelis-Menton Equation as a Basis for Assay Conditions

The Michaelis-Menton equation (see references *5* and *8* for the derivation) expresses the quantitative aspects of enzyme kinetics. For an enzyme reaction,

$$v_i = \frac{[S]v_{max}}{K_m + [S]} \tag{147}$$

where v_i and v_{max} are the initial and maximal velocities, respectively, of the reaction; $[S]$ is the substrate concentration at the time v_i is measured; and K_m is the Michaelis constant, defined as the substrate concentration at half maximal velocity. The equation relates the initial velocity of the reaction to the maximal velocity and substrate concentration. If the true initial velocities are measured, then $[S] = [S_0]$ where $[S_0]$ is the initial substrate concentration. A plot of v_i versus $[S_0]$ illustrates this relationship (Figure 6.1). In the figure, the kinetics in region **A** form the basis for conditions required by assays that determine an enzyme's activity. Likewise, the kinetics in region **B** form the basis for conditions required by assays that determine the concentration of a substrate or co-substrate, using an enzyme as a reagent.

B. Assays that Measure Enzyme Activity

In region **A** of Figure 6.1, v_i does not change as $[S_0]$ changes. When the rate of a reaction is independent of the concentrations of the reactants, the reaction has **zero-order** kinetics. Thus, reactions in region **A** have zero-order kinetics. Also, $[S_0] > K_m$, and the $[S]$ in the denominator of equation 147 is, therefore, far larger than K_m. The latter can then be ignored in the equation. In quantitative terms,

$$K_m + [S] \approx [S]$$

and equation 147 becomes:

$$v_i = \frac{[S]}{[S]} v_{max} = v_{max} \tag{148}$$

This means that in region **A**, the rate at which an enzyme converts its substrate to product (P) is maximal and constant. Consequently, as the reaction proceeds, $[S]$ decreases while $[P]$ increases linearly. When either is plotted against time (t), the slope of the curve yields the activity of the enzyme (Figure 6.2). In general, these properties of region **A** form

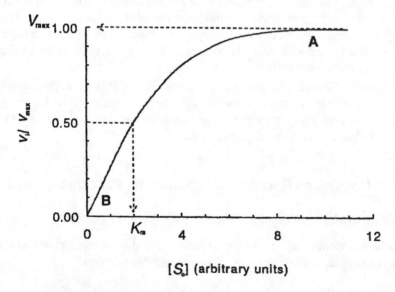

Figure 6.1. Relationship between initial velocity and initial substrate concentration of an enzyme-catalyzed reaction.

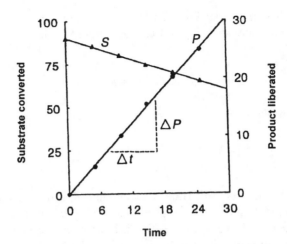

Figure 6.2. A plot of [*S*] or [*P*] versus time (*t*) when [*S*] is saturating. $\Delta P/\Delta t$ is a measure of the reaction's velocity (*v*).

the criteria for all quantitative assays that measure the optimal activities of enzymes. These criteria are summarized below:

(1) [*S*] should be saturating ([*S*] ≈ 100 × K_m is suggested).

(2) The enzyme concentration used should give a linear product formation or substrate consumption over a time period sufficiently long to permit making accurate measurements.

(3) **To verify that the above conditions are met, [*S*] or [*P*]** is plotted against *t*, and the curve should be linear. Alternatively, v_i is plotted against *t*, and this curve should be a straight line parallel to the time axis. Also, a doubling of the enzyme concentration should result in a doubling of v_i.

Example: See problem 102 in section 6.5.

(4) In addition to the above criteria specified by the Michaelis-Menton kinetics, the temperature and pH of the assay must be optimized because a given enzyme catalyzes optimally at a certain pH and temperature peak or within certain pH and temperature ranges.

The required enzyme concentration should be determined by assaying several dilutions of the enzyme using a fixed saturating [*S*] (see problem 102). If K_m is unknown, the saturating [*S*] should also be estimated by assaying several [*S*] using a fixed adequate enzyme concentration. When v_i is plotted against [S_0] in such a case, the saturating [S_0] values are those in which $v_i = v_{max}$. In Figure 6.1, for example, the saturating [*S*] values are those that are greater than about 4.5. To determine optimal temperature, a series of assays is performed at different temperatures. The activity for each assay is estimated and plotted against temperature. The optimal temperature corresponds to the highest activity without detrimental side effects . Optimal pH is determined similarly, except that the assay is performed at different pH values. Data from assays that meet the above conditions are then used to calculate the enzyme's activity and kinetic constants using equations given in sections 6.3 and 6.4.

C. *Assays that Measure Substrate Concentration Using the Enzyme as a Reagent*

In region **B** of Figure 6.1, v_i varies linearly with [*S*]. When the rate of a reaction is dependent on the concentration of one reactant, the reaction has **first-order kinetics**. Region **B**, therefore, has first-order kinetics. $[S] \ll K_m$, and the $[S]$ in the denominator of equation 147 is small relative to K_m and can be ignored in the equation. In quantitative terms,

$$K_m + [S] \approx K_m$$

and equation 147 becomes

$$v_i = \frac{v_{max}}{K_m}[S] = k[S] \tag{149}$$

where $k = (v_{max}/K_m)$ is the first-order rate constant, and $[S]$ is the substrate concentration at the time v_i is measured. Equation 149 implies that when substrate concentration is limiting (i.e., $[S] \ll K_m$), the rate of an enzyme reaction is linearly proportional to the substrate concentration, the proportionality constant being v_{max}/K_m. This relationship forms the basis for setting up assays to measure the concentrations of substrates, using enzymes as reagents. The assays require the following conditions:

(a) The substrate concentration should be limiting (i.e., $[S] \ll K_m$) so that first-order kinetics are established.

(b) The enzyme concentration used should give a linear product formation or substrate consumption over a time period sufficiently long to permit making accurate measurements.

Data from assays that meet these conditions are then used to calculate the substrates concentration, as explained below.

(1) Calculating $[S_0]$ based on kinetic data

The concentration to be determined corresponds to $[S_0]$, the substrates concentration in the assay medium at $t = 0$. An equation that relates $[S_0]$ to the $[S]$ measured at a later time (t) during the kinetic phase of the enzyme reaction is derived by using equation 149 and the fact that v_i is the initial rate of decrease in the substrate concentration as time changes:

$$v_i = \frac{d[S]}{dt} = k[S] \tag{150}$$

$$\therefore \quad -\frac{d[S]}{[S]} = kdt \tag{151}$$

The following equations are obtained when equation 151 is integrated over the limits: $[S_0]$ to $[S]$ (for the left side) and 0 to t (for the right side). See Appendix A.5 for details.

$$[S] = [S_0]e^{-kt} \tag{152}$$

$$\therefore [S_0] = \frac{[S]}{e^{-kt}} \tag{153}$$

Equations 152 and 153 are the general expressions for a first-order reaction and are used to calculate $[S_0]$, if k is known ($k = v_{max}/K_m$). Alternatively, by taking readings at two times, t_1 and t_2, during the reaction, the change in substrate concentration (ΔS) during this interval can be used to calculate $[S_0]$ according to equation 154, below. See Appendix A.5 for details.

$$[S_0] = \frac{\Delta[S]}{(e^{-kt_1} - e^{-kt_2})} \tag{154}$$

(2) Determining [S_0] based on end-point assay

With the end-point assay technique, the reaction is allowed to proceed to equilibrium so that practically all the substrate molecules have been converted to product before a reading is taken. Standards of the pure product are used to generate a standard curve from which the [S_0] of samples are read. The enzyme concentration in the assay should be sufficient for a complete conversion of the substrate within a reasonable assay time. If the equilibrium does not favor completion, the reaction can be coupled to a secondary reaction capable of displacing the equilibrium in favor of completion.

6.3.   Enzyme Activity and Units

The activity of an enzyme is defined with respect to the amount of substrate catalyzed under a given set of assay conditions. It is calculated using raw data obtained from the type of assays described in section 6.2B and the equations derived below.

A.   Activity

Activity is defined as the amount of substrate converted or product liberated per unit time under a defined set of assay conditions. Activity is calculated using equation 155 or 156, below:

$$\text{activity} = \frac{\text{amount of } S \text{ converted}}{t} \tag{155}$$

$$\text{or activity} = \frac{\text{amount of } P \text{ liberated}}{t} \tag{156}$$

Examples: See problems 100a and 102b in section 6.5.

B.   Specific Activity (sp act.)

Specific activity is the amount of substrate converted or product liberated per unit time per milligram of protein under a defined set of assay conditions. **Specific activity is calculated using equation 157, 158, or 161, below:**

$$\text{sp act.} = \frac{\text{amount of } S \text{ converted}}{t \times \text{mg of protein}} \tag{157}$$

$$\text{or sp act.} = \frac{\text{amount of } P \text{ liberated}}{t \times \text{mg of protein}} \tag{158}$$

The "mg of protein" in the above equations refers to the total milligrams of all proteins present in the sample, not just in the enzyme. Decreasing the amount of contaminating proteins during purification decreases the denominator of equation 157, resulting in an increase in specific activity. Therefore, as the purity of an enzyme preparation increases, the specific activity of the enzyme also increases.

Examples: See problems 97a, 98, 100, 101, and 102 in section 6.5.

C. Enzyme Activity Units

An activity unit allows one to estimate the quantity of enzyme that possesses a given activity. In general, one unit of activity for an enzyme is the amount of the enzyme that converts a given amount of substrate per unit time under optimal assay conditions.

(1) International Unit (IU)

This widely used unit is defined as follows: One **IU** of an enzyme is the amount that catalyzes the conversion of 1 μmol of substrate per minute under a defined set of assay conditions. When expressed mathematically,

$$1 \text{ IU} = 1 \text{ μmol/min} \tag{159}$$

The concentration of the enzyme can then be expressed as IU/unit volume of sample or IU/unit weight of proteins in the sample. To convert enzyme activity from other units to IU, first convert the units of S or P to μmol and the time to minutes. Use equation 160 or 161 below, to calculate the activity in IU.

$$\text{activity (IU)} = \frac{\text{μmol of } S \text{ or } P \text{ catalyzed}}{\text{min}} \tag{160}$$

$$\text{and sp. act. (IU)} = \frac{\text{μmol of } S \text{ or } P \text{ catalyzed}}{\text{min} \times \text{mg of protein}} \tag{161}$$

Examples: See problems 101 and 102b in section 6.5.

(2) Katal (kat)

This unit is defined as the amount of enzyme that converts 1 mole of substrate per second. Thus,

$$1 \text{ kat} = 1 \text{ mol/s} = 6 \times 10^7 \text{ μmol/min} = 6.0 \times 10^7 \text{ IU}$$

$$\therefore 1 \text{ IU} = 1/60 \text{ μkat}$$

6.4. Basic Kinetic Constants: K_{cat}, v_{max}, and K_m

Detailed discussions on enzyme kinetics can be found in references 6, 9, and 10. Only the kinetic constants, K_{cat} (turnover number), v_{max}, and K_m, are defined and briefly discussed below.

A. Turnover Number (K_{cat})

Turnover number or catalytic constant (K_{cat}), is the maximum number of moles of substrate converted to product per second per mole of active site on the enzyme. If there is one active site per enzyme molecule, then turnover number is the number of substrate molecules converted by one molecule of the enzyme per second. For example, a turnover number of

$500 \, s^{-1}$ means that each enzyme molecule converts 500 molecules of substrate to product each second when substrate concentration is saturating. From the above definition,

$$\text{turnover no.} = \frac{v_{max} (\text{mol} / \text{s})}{[\text{enzyme}](\text{mol})} = K_{cat} \qquad (162)$$

Rearrangement yields equation 163, which is used to calculate the turnover number of an enzyme.

$$\text{turnover no.} = \frac{\text{mol of S catalyzed}}{\text{s} \times \text{mol of enzyme}} \qquad (163)$$

Example: See problem 100c in section 6.5.

B. Maximal Velocity (V_{max})

This is the maximal velocity attained by a enzyme reaction when substrate concentration is saturating. At v_{max}, all the enzyme molecules are complexed with substrate molecules, and as soon as the bound substrate is converted and released, another one moves in for another round of turnover. v_{max} is a constant only when substrate concentration is saturating; its calculation is explained below.

C. Michaelis Constant (K_m)

The K_m for an enzyme is the substrate concentration at which the rate of the enzyme's reaction is one-half of v_{max}. It is a constant characteristic of a given enzyme. The lower the K_m, the higher the affinity of the enzyme to the substrate, and when alternative substrates exist, the one with the lowest K_m fits best into the enzyme's active site. Knowledge of K_m facilitates (1) estimating the *in vivo* or *in vitro* optimum substrate concentration, (2) establishing how an enzyme is regulated, and (3) identifying differences in an enzyme isolated from different sources (6).

v_{max} and K_m are readily obtained from a plot of v_i versus $[S_0]$ (see Figure 6.1). More accurate values are obtained using a Lineweaver-Burk double reciprocal plot (Figure 6.3). The latter is a plot of equation 164, below, which is the inverse of equation 147:

$$\frac{1}{v_i} = \frac{K_m}{v_{max}} \times \frac{1}{[S_0]} + \frac{1}{v_{max}} \qquad (164)$$

This being a linear equation means that

$$y \text{ intercept} = \frac{1}{v_{max}}$$

$$\therefore v_{max} = \frac{1}{y \text{ intercept}} \qquad (165)$$

$$x \text{ intercept} = \frac{1}{K_m}$$

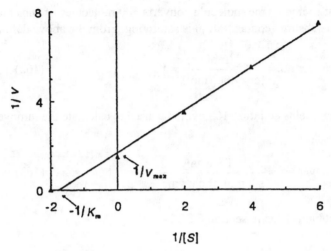

Figure 6.3. A Lineweaver-Burk double reciprocal plot (equation 164).

$$\therefore K_m = -\frac{1}{x \text{ intercept}} \tag{166}$$

$$\text{and slope} = \frac{K_m}{v_{max}} \tag{167}$$

6.5. Practical Examples

PROBLEM 100

A 0.10 mL aliquot of a pure ß-galactosidase (ß-gal) solution (1.00 mg protein/mL) hydrolyzed 0.10 mmol of *o*-nitrophenyl galactoside (ONPG) in 5.0 min. Calculate (a) the activity, (b) the specific activity, and (c) the turnover number of the ß-gal. Assume a mol wt of 116.0 kD for the enzyme and one active site per molecule.

SOLUTION:

(a) Convert the amount of ß-gal hydrolyzed to µmoles, and the time (*t*) to minutes.

$$\text{amount of ß-gal hydrolyzed} = 0.10 \text{ mmol} \times 1000 \text{ (µmol/mmol)}$$
$$= 100.0 \text{ µmol}$$
$$t = 5.0 \text{ min}$$

Using equation 160,

$$\text{activity (IU)} = \frac{\text{µmol of ONPG hydrolyzed}}{\text{min}}$$

$$= \frac{100.0 \text{ µmol}}{5.0 \text{ min}} = 20.0 \text{ µmol / min} = 20.0 \text{ IU}$$

where 1 IU = µmol/min.

(b) Total protein in assay = 1.00 mg/mL × 0.10 mL = 0.100 mg. From equation 161,

$$\text{sp act.} = \frac{\mu\text{mol of ONPG hydrolyzed}}{\text{min} \times \text{mg of protein}}$$

$$= \frac{100.0 \ \mu\text{mol}}{5.0 \ \text{min} \times 0.10 \ \text{mg of protein}}$$

$$= 200 \ \mu\text{mol min}^{-1} \ \text{mg}^{-1} \ \text{of protein}$$

$$= 2.0 \times 10^2 \ \text{IU/mg of protein}$$

(c) Using equation 1,

$$\text{mol of enzyme in assay medium} = \frac{0.00010 \ \text{g}}{116,000 \ \text{g / mol}} = 8.6 \times 10^{-10} \ \text{mol}$$

$$\text{mol of ONPG hydrolyzed} = \frac{0.10 \ \text{mmol}}{1000 \ \text{mmol / mol}} = 1.0 \times 10^{-4} \ \text{mol}$$

Using equation 163,

$$K_{\text{cat}} \text{ turnover number} = \frac{\text{mol of ONPG hydrolyzed}}{\text{s} \times \text{mol of enzyme}}$$

$$= \frac{1.0 \times 10^{-4} \ \text{mol}}{5.0 \ \text{min} \times 60 \ (\text{s / min}) \times 8.6 \times 10^{-10} \ \text{mol}} = 390 \ \text{s}^{-1}$$

PROBLEM 101

A vial of lyophilized hexokinase (total protein, 100.00 mg; sp act. 200 IU/mg of protein) is reconstituted with 10.0 mL of diluent. (a) Calculate the units/mL of hexokinase in the solution. (b) What volume of this solution will yield 50 IU of hexokinase per mL of assay medium? (c) Repeat the calculations for a vial that contains 20,000 IU of total activity and is reconstituted with 5.0 mL of diluent.

SOLUTION:

(a)

$$\text{Units / mL} = \frac{\text{total activity}}{\text{total volume}}$$

$$= \frac{200 \ \text{IU / mg of protein} \times 100.00 \ \text{mg of protein}}{10.0 \ \text{mL}} = 2,000 \ \text{IU / mL}$$

(b) Let V_1 be the volume of hexokinase solution needed to prepare the 1 mL (V_2) solution; C_1 and C_2 are the hexokinase concentrations in the stock and final solutions, respectively. By using equation 39,

$$V_1 = \frac{C_2 V_2}{C_1} = \frac{50 \ \mu\text{mol min}^{-1}\text{mL}^{-1} \times 1 \ \text{mL}}{2,000 \ \mu\text{mol min}^{-1}\text{mL}^{-1}}$$

$$= 0.02 \text{ mL or } 20.0 \text{ μL}$$

∴ **Twenty μL of the hexokinase solution is added to 0.980 mL of assay medium to yield 50 IU/mL.**

(c) Calculate the hexokinase activity/mL of solution

$$\text{units / mL} = \frac{\text{total activity}}{\text{total volume}} = \frac{20{,}000 \text{ IU}}{5.0 \text{ mL}} = 4{,}000 \text{ IU/mL}$$

Following the example in (b),

$$V_1 = \frac{C_2 V_2}{C_1} \quad \frac{50 \text{ IU / mL} \times 1.0 \text{ mL}}{4{,}000 \text{ IU / mL}} = 0.01 \text{ mL or } 10.0 \text{ μL}$$

where C and V are defined as in (b), above.

Therefore, 10.0 μL of the hexokinase solution is added to 990.0 μL of the assay medium to yield 50 IU/mL.

PROBLEM 102

Alcohol dehydrogenase in a cell extract was assayed by adding 0.1 mL of the undiluted or diluted extract to 1.9 mL of the assay medium containing ethanol (EtOH) and NAD^+ in a 1.0 cm cuvette. The absorbance at 340 nm was monitored for 10.0 min and the data obtained are shown in Table 6.1, below. (a) Which data should be used to calculate the activity of alcohol dehydrogenase and why? (b) Calculate the total activity of alcohol dehydrogenase in 20.0 mL of the extract. The E_{1cm}^{1M} of NADH at 340 nm is 6220.

SOLUTION:

(a) To determine which set of data should be used to calculate the activity, plots of $[P]$ versus t (Figure 6.4) and v versus t (Figure 6.5) are made. Data sets (1) and (2) should not be used to calculate the activity because the $[P]$ versus t plot is not linear and the v versus t plot is not constant for either set. Zero-order kinetics is required for assays to measure activity, in which case, v must be constant and $[P]$ must increase linearly with time (see section 6.2.B). These conditions are satisfied by data sets (3) and (4) (Figures 6.4 and 6.5); hence, either set can be used to calculate the activity. However, the reaction for data set (4) is somewhat too slow. Data set (3) is more appropriate for calculating the activity.

(b) Any of the intervals within the linear phase of the reaction can be used for the calculations. For the interval between 0 and 8 min,

Table 6.1. Data for Problem 102

| OD | Time (min) | | | | | | DF |
	0.0	2.0	4.0	6.0	8.0	10.0	
(1)	0	0.75	0.97	1.04	1.06	1.09	1
(2)	0	0.40	0.70	0.90	0.97	1.04	3
(3)	0	0.12	0.24	0.36	0.48	0.60	10
(4)	0	0.06	0.16	0.18	0.24	0.30	20

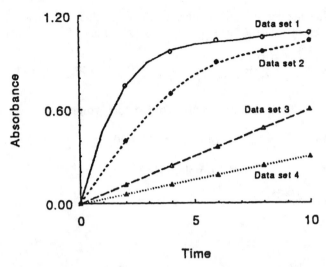

Figure 6.4. Plots of product concentration [P] versus time for the data sets given in problem 102. [P] is approximated by OD.

$$t = (8.0 - 0.0) \text{ min} = 8.0 \text{ min}; \text{ OD} = (0.48 - 0) = 0.48.$$

First, using equation 143, calculate the amount of NADH formed:

$$[\text{NADH}] = \frac{A}{E\ell} = \frac{0.48}{6220 \text{ M}^{-1}\text{cm}^{-1} \times 1.0 \text{ cm}} = 7.72 \times 10^{-5} \text{ M}$$

mole of NADH formed in 2.0 mL assay medium

$$= 7.72 \times 10^{-5} \text{ (mol / L)} \times \frac{2.0 \text{ mL}}{1000 \text{ (mL / L)}} = 1.54 \times 10^{-7} \text{ mol} = 0.154 \text{ μmol}$$

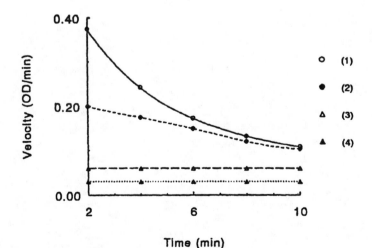

Figure 6.5. Plots of velocity (OD/min) versus time for the data given in problem 102.

Second, calculate the activity/mL of the extract using equation 156, and taking into account the dilution factor 10 and the fact that only 0.10 mL of the 20.0 mL extract was assayed.

$$\text{activity} / \text{mL} = \frac{0.154 \ \mu\text{mol}}{10.0 \ \text{min}} \times \frac{1}{0.10 \ \text{mL}} \times 10 = 1.54 \ \mu\text{mol min}^{-1} \ \text{mL}^{-1}$$

$$\text{Total activity} = \text{activity/mL} \times \text{total volume}$$

$$= 1.54 \ \mu\text{mol min}^{-1} \ \text{mL}^{-1} \times 20.0 \ \text{mL}$$

$$= 31 \ \mu\text{mol/min} = 31 \ \text{IU}$$

where 1 IU = 1 μmol/min.

Chapter 7

Radioactivity and Related Calculations

7.1. Introduction

The isotopes of an element contain the same number of protons but different numbers of neutrons in the atomic nucleus. They also have the same number of electrons. Thus, they have identical chemical properties because the number of electrons is what determines the chemical properties of the element. Whether or not an isotope is stable depends on the relative numbers of protons and neutrons in its nucleus. When an isotope is unstable, it disintegrates to another isotope of the same or different element by emitting α, ß, γ, or other particles. Such an isotope is termed radioactive, and those that emit ß or γ particles are widely used in biological work.

A ß particle has a continuous range of energy values characteristic of the emitting isotope, and this is exploited in liquid scintillation to discriminate between different isotopes in a sample. A γ ray is a photon and has discrete energy values, in contrast to the continuous energy spectrum of the ß particle. A given isotope emits γ photons of one or more discrete energy values characteristic of the isotope. This is also exploited to discriminate between different γ-emitting isotopes in a sample. To measure the amount of radioactivity, either liquid scintillation or the Geiger-Müller counting technique is used to measure the rate of particle emission. The underlying principles for these techniques are briefly discussed in Appendix A.6.

Applications of radioactivity to studies in the life sciences consist of labeling biomolecules with atoms of a radioactive element such as ^{3}H, ^{14}C, ^{32}P, or ^{35}S, and using the labeled material to assay various cellular and biochemical functions. The quantity of radioactivity used is critical to the success of the assay, and the quantity recovered enables the experimenter to transform raw data to desired results. Basic calculations for experiments involving radioactivity are presented in this chapter.

7.2. Calculations Involving Radioactivity

Basic calculations involving radioactivity are performed on the basis of the quantitative descriptions of radioactive decay and the definitions of radioactivity units and specific radioactivity. These are discussed below.

A. Quantitative Description of Radioactive Decay

For a given radioactive substance, the number of decaying atoms (dN) per unit time interval (dt) is proportional to the initial number of radioactive atoms (N_0). When expressed mathematically,

$$-\frac{dN}{dt} = \lambda N_0 \tag{168}$$

where λ is the proportionality constant. Both sides of equation 168 are then integrated from N_0 to N_t (left side), and from 0 to t (right side); N_0 and N_t being the number of radioactive atoms present at $t = 0$, and at a later time, t, respectively:

$$-\int_{N_0}^{N_t} \frac{dN}{N_0} = \int_0^t \lambda dt \tag{169}$$

$$\ln \frac{N_0}{N_t} = \lambda t \tag{170}$$

Taking the antilog,

$$N_t = N_0 e^{-\lambda t} \tag{171}$$

Because the value of λ is not always available, equation 171 is modified so that a more readily available constant, $t_{1/2}$ **(half life)** can be used to calculate N_t. The $t_{1/2}$ is defined as the time it takes one-half of a given radioactive material to decay. If $N_0 = 1$, then at $t_{1/2}$, $N_t = 0.5$. Substituting these into equation 171,

$$0.5 = e^{-\lambda t_{1/2}}$$

$$\ln 0.5 = -\lambda t_{1/2} \ln e = -\lambda t_{1/2}$$

$$\therefore 1 = \frac{-\ln 0.5}{t_{1/2}} = \frac{0.693}{t_{1/2}}$$

Substituting this expression for in equation 171 yields

$$N_t = N_0 e^{-0.693 t / t_{1/2}} \tag{172}$$

In practical terms, N_t is the amount of radioactivity remaining in a sample after a given time has elapsed, and is calculated using equation 172.

Example: See problem 105 in section 7.3.

B. Units of Radioactivity

Common units of radioactivity are listed in Table 7.1 along with their equivalents in other units. Many calculations relating to radioactivity can be done based on these definitions.

Table 7.1. Units of Radioactivity and Their Equivalents

Unit	Definition	Equivalent
Curie (Ci)	The amount of a radioactive substance decaying at a rate of 3.7×10^{10} disintegrations per second (dps).	1 Ci $= 3.7 \times 10^{10}$ dps $= 2.22 \times 10^{12}$ dpm
Microcurie (μCi)	One-millionth of a curie.	1 μCi $= 2.22 \times 10^{6}$ dpm
Disintegrations per minute (dpm)	The number of radioactive atoms disintegrating per minute.	
Counts per minute (cpm)	The number of disintegrations detected per minute. If the counting device is 100% efficient, then cpm is equal to dpm.	cpm = dpm × counting efficiency
Becquerel (Bq)	The SI unit of radioactivity defined as the quantity of a radioactive substance decaying at a rate of 1 dps.	1 Bq = 1 dps = 60 dpm $= 2.7 \times 10^{-11}$ Ci 1 Ci $= 3.7 \times 10^{10}$ Bq 1 mCi $= 3.7 \times 10^{7}$ Bq 1 μCi $= 3.7 \times 10^{4}$ Bq

Examples: See problems 103 and 104 in section 7.3.

C. Specific Radioactivity and the Calculation of Concentrations

In a labeling process, atoms of a specific element in a molecule are randomly replaced by radioactive atoms of the same element. Replacing too many atoms can adversely affect the function of the molecule because of high radiation effects; replacing too few can affect the sensitivity of the assay because the radiation signal would be too low for accurate detection. In practice, the proper amount is determined experimentally for particular molecules. To enable the experimenter to determine what quantity of the labeled material contains a given count rate and vice versa, the radioactivity is expressed per unit amount of the labeled material. **The quantity of radioactivity per unit amount of labeled material is termed specific radioactivity or, simply, specific activity.** Examples of specific radioactivity units include μCi/mg, mCi/μmol, cpm/mol, μCi/mL, and Bq/mg. To calculate specific activity, the following equation is used:

$$\text{sp radioactivity} = \frac{\text{radioactivity of a sample}}{\text{amount of the sample}} \qquad (173)$$

Specific radioactivity can be used to calculate the amount of a radioactive substance needed to produce a given count rate, the concentration of a biological receptor that has been labeled with a radioactive ligand, or the intracellular concentration of solutes. These applications are best illustrated by specific **examples** such as problems 104, 105, 106, 107, and 108.

7.3. Practical Examples

PROBLEM 103

Convert 1.0×10^{7} dpm to (a) μCi and (b) cpm. Assume a counting efficiency of 80%.

SOLUTION:
(a) From Table 7.1, 2.22×10^6 dpm = 1 μCi

$$\therefore \text{ the number of μCi in } 1.0 \times 10^7 \text{ dpm}$$

$$= \frac{1 \text{ μCI}}{2.22 \times 10^6 \text{ dpm}} \times 1.0 \times 10^7 \text{ dpm}$$

$$= 4.5 \text{ μCi}$$

(b) From Table 7.1, cpm = dpm × counting efficiency

$$-1.0 \times 10^7 \text{ dpm} \times \frac{80 \text{ cpm}}{100 \text{ dpm}} = 8.0 \times 10^6 \text{ cpm}$$

PROBLEM 104

(a) Starting with solid ^{14}C-inulin (sp act. 100.0 μCi/mg), prepare a 10.0 mL solution containing 500.0 μCi of radioactivity. (b) What volume of this solution will produce a count rate of 1.0×10^6 cpm in a 2.0 mL assay medium? (c) If 0.2 mL of the assay mixture is counted, what cpm is expected? Assume an 80% counting efficiency.

SOLUTION:
(a) The total μCi in the 10.0 mL solution = 500 μCi.
From the given specific activity, 100.0 μCi is contained in 1.0 mg of the ^{14}C-inulin.

$$\therefore 500.0 \text{ μCi is contained in } \frac{1 \text{ mg}}{100.0 \text{ μCi}} \times 500.0 \text{ μCi} = 5.0 \text{ mg}$$

To prepare the solution, weigh 5.00 mg of the ^{14}C-inulin and dissolve it such that the final volume is 10.0 mL. This solution contains a total of 500.0 μCi of radioactivity.

(b) Convert the count rate in the solution to cpm before calculating the required volume. From Table 7.1, 1 μCi contains 2.22×10^6 dpm.

$$\therefore 500.0 \text{ μCi contains } \frac{2.22 \times 10^6 \text{ dpm}}{1 \text{ Ci}} \times 500.0 \text{ μCi}$$

$$= 1.11 \times 10^9 \text{ dpm}$$

$$= 1.11 \times 10^9 \times 80\%/100\% = 8.88 \times 10^8 \text{ cpm}$$

This 8.88×10^8 cpm is contained in 10.0 mL.

$$\therefore \text{ The } 1.0 \times 10^6 \text{ cpm is contained in}$$

$$\frac{10.0 \text{ mL}}{8.88 \times 10^8 \text{ cpm}} \times 1.0 \times 10^6 \text{ cpm} = 0.011 \text{ mL or } 11 \text{ μL}$$

A solution volume of 11 μL contains the required 1.0×10^6 cpm in the assay medium.
(c) 2.0 mL of assay medium contains 1.0×10^6 cpm.

$$\therefore \text{ The 0.2 mL contains } \frac{1.0 \times 10^6 \text{ cpm}}{2.0 \text{ mL}} \times 0.2 \text{ mL} = 1.0 \times 10^5 \text{ cpm}$$

PROBLEM 105

A purchased ^{32}P-labeled sample (4.5×10^7 cpm/pmol) could not be used until 6 days after its reference date. Calculate the specific activity remaining in the sample. The $t_{\frac{1}{2}}$ of ^{32}P is 14.2 days.

SOLUTION:

Let the radioactivity remaining be N_t. Then, using equation 172,

$$N_t = N_0 e^{-0.693t/t_{\frac{1}{2}}}$$

$$= 4.5 \times 10^7 \text{ cpm/pmol} \times e^{(-0.693 \times 6 \text{ days})/14.2 \text{ days}}$$

$$= 3.4 \times 10^7 \text{ cpm/pmol}$$

Note: The units for t and $t_{\frac{1}{2}}$ must be the same.

PROBLEM 106

In a binding assay, labeled hormone (5.0×10^5 cpm/mg) was added to 1.0 mL of the assay medium containing the hormone's receptor. A 0.1 mL aliquot was then processed and counted for bound hormone. If the count rate was 1.0×10^4 cpm, calculate the total amount of hormone bound in the 1.0 mL of assay medium. The receptor has 1 binding site per molecule.

SOLUTION:

First, calculate the total bound cpm in the 1.0 mL of assay medium:

$$1.0 \times 10^4 \text{ cpm/0.1 mL} = 1.0 \times 10^5 \text{ cpm/mL}$$

Second, calculate the milligrams of hormone containing the 1.0×10^5 cpm. From the given specific activity, 5.0×10^5 cpm is contained in 1 mg of hormone.

$$\therefore 1.0 \times 10^5 \text{ cpm is contained in } \frac{1 \text{ mg}}{5.0 \times 10^5 \text{ cpm}} \times 1.0 \times 10^5 \text{ cpm} = 0.20 \text{ mg}$$

$$\therefore \text{ bound hormone} = 0.20 \text{ mg}$$

PROBLEM 107

In a Na^+ transport assay, a total of 1.0×10^6 cpm of ^{22}NaCl (sp act., 4.0×10^8 cpm/μg) was added to 1.0 mL of the assay medium containing 1.5 mM non-radioactive NaCl and bacterium vesicles. (a) Calculate the specific activity of ^{22}Na$^+$ in the medium. (b) If vesicles in 0.1 mL of the medium accumulated 5.0×10^3 cpm of the ^{22}Na$^+$, calculate the moles of Na$^+$ accumulated.

SOLUTION:

The amount of Na$^+$ contributed by the ^{22}NaCl (less than 50 nmol) is negligible compared to contribution from the non-radioactive NaCl.

(a) total amount of Na$^+$ = 1.5 (mmol/L) × 0.001 L = 1.5 μmol

$$\text{total radioactivity} = 1.0 \times 10^6 \text{ cpm}$$

$$\therefore \text{sp act.} = \frac{1.0 \times 10^6 \text{ cpm}}{1.5 \text{ } \mu\text{mol}} = 6.7 \times 10^5 \text{ cpm} / \mu\text{mol}$$

(b) 6.67×10^5 cpm is contained in 1 μmol of Na^+ in the medium.

$$\therefore 5.0 \times 10^3 \text{ cpm is contained in}$$

$$\frac{1 \text{ } \mu\text{mol Na}^+}{6.67 \times 10^5 \text{ cpm}} = 5.0 \times 10^3 \text{ cpm} = 0.0075 \text{ } \mu\text{mol Na}^+$$

In the 0.1 mL aliquot, 0.0075 μmol Na^+ (or 7.5 nmol Na^+) were accumulated by vesicles.

PROBLEM 108
A 50-base, single-strand synthetic DNA was purified and dissolved in 0.5 mL of buffer. If a 1 to 100 dilution of it has an OD of 0.25, (a) calculate the total moles of DNA in the solution. (b) If 10.0 pmol of the DNA is needed in a 20.0 μL assay mixture, what volume of the DNA solution should be added? (c) The 10.0 pmol aliquot was labeled with ^{32}P, and 70% was recovered and re-dissolved in 100.0 μL buffer. If 2.0 μL had a count rate of 450,500 cpm, calculate the specific activity of the labeled DNA.

SOLUTION:
(a) Using equation 146 and AU values from Table 6.1,

$$\text{concentration of DNA} = 1 \text{ AU} \times \text{OD} \times \text{DF} = 40 \text{ } \mu\text{g/mL} \times 0.655 \times 100 = 2620 \text{ } \mu\text{g/mL}$$

$$\text{Total DNA} = 2620 \text{ } \mu\text{g/mL} \times 0.5 \text{ mL} = 1310.0 \text{ } \mu\text{g}$$

$$\text{Molecular mass of DNA} = 650 \text{ D/bp} \times 50 \text{ bases} = 32,500 \text{ D.}$$

Using equation 3,

$$\text{mol DNA} = \frac{\text{wt}}{\text{mol wt}} = \frac{1.31 \times 10^{-3} \text{ g}}{32,500 \text{ g / mol}} = 4.03 \times 10^{-8} \text{ mol}$$

$$= 4.03 \times 10^{-8} \text{ mol} \times \frac{1 \text{ pmol}}{1 \times 10^{-12} \text{ mol}} = 4.0 \times 10^4 \text{ pmol}$$

(b) The concentration of DNA can be expressed as follows:

$$\frac{40,300 \text{ pmol}}{500.0 \text{ } \mu\text{L}} = 80.6 \text{ pmol} / \mu\text{L}$$

$$\therefore \text{volume containing 10.0 pmol is } \frac{1 \text{ } \mu\text{L}}{80.6 \text{ pmol}} \times 10.0 \text{ pmol} = 0.12 \text{ } \mu\text{L}$$

This volume is too small to pipette accurately with a micropipettor. Therefore, dilute an aliquot of the sample to a final concentration of 2.0 pmol/μL. 5.0 μL of this solution contains 10.0 pmol of DNA. Use equation 39 to calculate V_1, the volume of the stock DNA solution that will yield 2.0 pmol/µL (C_2) in a final volume of 1.0 mL (V_2).

$$V_1 = \frac{C_2 V_2}{C_1} = \frac{2.0 \text{ pmol} / \mu L \times 1 \text{ } \mu L}{100.0 \text{ pmol} / \mu L} = 20.0 \text{ } \mu L$$

Therefore, add 20.0 µL of the stock DNA solution to 980.0 µL of diluent to obtain a solution of 2.0 pmol/µL. Add 5.0 µL of this solution to the reaction mixture.

(c) Since the recovery was 70%,
Amount of labeled DNA recovered

$$\frac{70\%}{100\%} \times 10.0 \text{ pmol} = 7.0 \text{ pmol}$$

∴ the amount of DNA/2.0 µL aliquot

$$\frac{7.0 \text{ pmol}}{100 \text{ } \mu L} \times 2.0 \text{ } \mu L = 0.14 \text{ pmol}$$

Radioactivity/2.0 µL aliquot = 450,500 cpm. Using equation 172,

$$\text{sp radioactivity} = \frac{\text{radioactivity of sample}}{\text{amount of sample}}$$

$$= \frac{450,500 \text{ cpm}}{0.14 \text{ pmol}} = 3.2 \times 10^6 \text{ cpm/pmol}$$

Chapter 8

Analytical Calculations Using Spreadsheets and Reporting Results into Laboratory Information Management Systems

by Todd Jones*

Today's laboratories are undergoing transformation to computer automation. Personal computers and the abundance of software programs designed specifically for the chemical and biological sciences are expanding the capabilities of modern laboratories, and concurrently forcing technical staff to be knowledgeable in applying basic chemical and biological calculations to a variety of tasks using computing tools. This chapter expands on the principles illustrated in previous chapters by demonstrating the power and convenience of using spreadsheets in such calculations. An overview of spreadsheet functionality and available software is presented. Six detailed examples of spreadsheet calculations are illustrated, including a full reference to spreadsheet content in Appendix M. Application of spreadsheets to Laboratory Information Management Systems (LIMS) is discussed in general terms at the end of the chapter.

8.1. Laboratory Automation

The advent of faster and cheaper personal computers is changing the landscape of chemical and biological laboratories. Computers predominantly control laboratory instrumentation. Laboratory communication is generated and disseminated using computers. Computers are handling laboratory sample information management. It is no longer sufficient to simply understand basic calculations in chemical and biological experiments; applying these basic calculations utilizing computer tools is a fundamental ability of laboratory personnel.

More and more laboratories are utilizing LIMS to control the flow of samples through various aspects of a business operation and laboratory analysis scheme. LIMS have the capability of tracking all data input, performing simple calculations, creating aliquot portions for special analyses, controlling instrument parameters and operations that generate analytical data, producing data reports, and auditing all aspects of a chemical or biological process. Using spreadsheets can enhance the database capabilities of LIMS by facilitating electronic transfer of sample data.

---

*Texas A&M University, College Station, TX 77843.

A spreadsheet is an electronic representation of a large sheet of paper with columns and rows. It is a graphical, visible, and interactive calculator. The spreadsheet organizes various sources of information, or data, into manageable and meaningful knowledge, to which some simple or complex calculation can be applied, producing even more information about the system being questioned. A spreadsheet can be used for the preparation of accounting records and financial statements, mathematical modeling, database conversions, or statistical analysis. The power of spreadsheets and LIMS together is enormous. In some cases, it can completely eliminate the manual recording of data. Besides performing calculations, spreadsheets can prevent the generation of erroneous results due to incorrect data input, perform extensive error detection, format data for clear presentation, organize extensive data observation from multiple experiments, and generate universally-readable files for import into other software programs. In chemical and biological applications, spreadsheets are best suited as "worksheets" or "templates" for routine or often repeated tasks. It takes time to create spreadsheets and to validate their internal calculations, but the time saved utilizing these templates over time is tremendous.

There are a number of spreadsheet software applications on the market. Unfortunately the software industry is constantly changing, releasing an endless stream of updates to current products and creating new software products. The programs listed below have been in the market place for years, in one version or another (*20*).

- Microsoft® Excel

- Corel® Quattro™ and Quattro Pro™

- Lotus® 1-2-3™

Each can generate the examples within this chapter and have similar features and functionality. The information in this chapter is generalized as much as possible to extend its applicability to multiple spreadsheet applications and multiple versions of these software products. The examples and methods contained in this chapter are for illustrative purposes only and were produced using Microsoft® Excel 97, version SR-2. For instructions or help using basic spreadsheet commands and functions, refer to the tutorial and manual provided with the software product.

8.2.   Analytical Calculations Using Spreadsheets

A.   Symbols Used in Examples

Spreadsheets utilize cells, or grid locations, to store information, calculations, or references to other information. Cell designations are given in the examples below as bolded-brackets **[X#]**, where X represents the column, and # represents the row. If only a column is referenced, a dash will be inserted in the # position (**[C-]** for column C). If only a row is referenced, a dash will be inserted in the X position (**[-3]** for row 3). A range of cells is represented by the first and last cell in the range, separated by a colon **[A1:B4]**. Formula expressions within cells are presented in distinguishing font type.

B.   Molecular Calculator

Determining the molecular mass (or molecular weight, mol wt, sections 1.1F and 2.2A) and percent composition (Section 2.2B) of a molecule is the first application of a spread-

	A	B	C	D	E	F
1						
2		**Molecular Weight/Weight Percent Determination**				
3						
4		Target Molecule:		$C_{12}H_{18}Cl_2N_4OS$		
5		Molecule Name:		Vitamin B_1-Thiamine hydrochloride		
6						
7	Atom	Atomic Weight	Quantity	Atom Total Weight	Weight Percent (%)	
8	H	1.00794	18	18.14292	5.38	
9	Li	6.941				
10	Be	9.01218				
11	B	10.811				
12	C	12.011	12	144.132	42.73	
13	N	14.00674	4	56.02696	16.61	
14	O	15.9994	1	15.9994	4.74	
15	F	18.9984				
16	Na	22.98977				
17	Mg	24.305				
18	Al	26.98154				
19	Si	28.0855				
20	P	30.9736				
21	S	32.066	1	32.066	9.51	
22	Cl	35.4527	2	70.9054	21.02	
23	K	39.0983				
24	Ca	40.078				
25						
26		TOTALS:	38	337.27268	100	
27				**Molar Mass (g/mol)**		
28						

Figure 8.1.

sheet calculation. Refer to the appropriate sections for the theoretical treatment and equations. A "molecular calculator," illustrated in Figure 8.1, is used to determine the molar mass and weight percent of the individual atoms within a molecule.

This spreadsheet consists of three sections. In the top section, the user can enter the target molecular formula and name. The middle section contains a partial list of elements in the Periodic Table most commonly found in chemical and biological applications, and their atomic weights. The bottom section contains a summation of the user information.

To use this spreadsheet, the user enters the target molecular formula **[C4:E4]**, name **[C5:E5]**, and the quantity of the individual atoms within the molecule **[C8:C24]**. The total weight of a particular atom within the molecule is calculated in cells **[D8:D24]** by multiplying the quantity by the Atomic Weight. A sum of the individual "Atom Total Weight" is designated as the Molar Mass in cell **[D26]**, with units of g/mol. The Weigh Percent of a particular atom is calculated in cells **[E8:E24]** by dividing the 'Atom Total Weight' by the Molar Mass, and multiplying this result by 100 (section 2.2B).

There are two cells for the user to check for gross errors. The sum of the atom quantities in cells **[C8:C24]** should equal the total atoms represented in the molecular formula in cell **[C4]**. The sum of each Weight Percent in cells **[E8:E24]** should equal 100%.

The cell formula under "Atom Total Weight" is expressed as:

$$=IF(\$C8="","",B8*C8)$$

	A	B	C	D	E	F
1						
2		Molecular Weight/Weight Percent Determination				
3						
4		Target Molecule:		$C_{216}H_{288}O_{144}$		
5		Molecule Name:		Starch triacetate		
6						
7	Atom	Atomic Weight	Quantity	Atom Total Weight	Weight Percent (%)	
8	H	1.00794	288	290.28672	5.59	
9	Li	6.941				
10	Be	9.01218				
11	B	10.811				
12	C	12.011	216	2594.376	50.00	
13	N	14.00674				
14	O	15.9994	144	2303.9136	44.40	
15	F	18.9984				
16	Na	22.98977				
17	Mg	24.305				
18	Al	26.98154				
19	Si	28.0855				
20	P	30.9736				
21	S	32.066				
22	Cl	35.4527				
23	K	39.0983				
24	Ca	40.078				
25						
26		TOTALS:	648	5188.57632	100	
27				Molar Mass (g/mol)		
28						
29			REAGENT PREPARATION			
30		Desired Concentration (M):		0.125		
31		Final Volume (L):		0.75		
32		Solid Reagent Needed (g):		486.4290		
33						

Figure 8.2.

Use the IF statement to check for an entry in column [C-]. If no user entry is found [=IF($C8=""], then the cell content [D-] is displayed as a blank (""), otherwise there is an entry in the Quantity cell, and the multiplication is performed [B8*C8]. Similarly, if no entry is found in the Quantity cell, then no weight percent should be calculated. This is represented by the cell formula:

$$=IF(C8="","",(D8/\$D\$31*100))$$

This calculation is performed, referencing an absolute cell [$D$31]. The $ symbol locks the cell location, and prevents it from being changed when cells are moved or copied to another location.

There is a limitation to all spreadsheet applications. The spreadsheet produces a result for any calculation, but will not properly round the result to the correct significant digits. It is up to the user to assess the proper number of significant digits to use, depending on whether this spreadsheet-generated result is used in further calculations, or as a final answer for a particular exercise.

Figures 8.2 and 8.3 illustrate two more examples of an organic and an inorganic molecule, respectively. Additionally, these two figures have a section "Reagent Preparation" that are utilized in the next example.

	A	B	C	D	E	F
1						
2		**Molecular Weight/Weight Percent Determination**				
3						
4		Target Molecule:		$LiHCa(CO_3)_2SO_4 \cdot 4H_2O$		
5		Molecule Name:		Fictitious Inorganic Cluster		
6						
7	Atom	Atomic Weight	Quantity	Atom Total Weight	Weight Percent (%)	
8	H	1.00794	9	9.07146	2.70	
9	Li	6.941	1	6.941	2.06	
10	Be	9.01218				
11	B	10.811				
12	C	12.011	2	24.022	7.15	
13	N	14.00674				
14	O	15.9994	14	223.9916	66.63	
15	F	18.9984				
16	Na	22.98977				
17	Mg	24.305				
18	Al	26.98154				
19	Si	28.0855				
20	P	30.9736				
21	S	32.066	1	32.066	9.54	
22	Cl	35.4527				
23	K	39.0983				
24	Ca	40.078	1	40.078	11.92	
25						
26		TOTALS:	28	336.17006	100	
27				**Molar Mass (g/mol)**		
28						
29			**REAGENT PREPARATION**			
30		Desired Concentration (M):		0.125		
31		Final Volume (L):		0.75		
32		Solid Reagent Needed (g):		31.5159		
33						

Figure 8.3.

Table M1 in Appendix M lists the cell contents for the spreadsheet template used in Figures 8.1, 8.2, and 8.3.

Examples: See problems 109, 110a, 111a, and 112.

C. Reagent Preparation Guide

The addition of a Regent Preparation Guide increases the usability of the Molecular Calculator spreadsheet. Throughout the text a simple routine calculation is used to determine the grams of a solid reagent needed to prepare a given solution. Section 1.2A describes the theoretical development.

Figures 8.2 and 8.3 have an added section "REAGENT PREPARATION." Under this header are two cells where the user can enter the desired concentration (in molarity) and the final volume (in liters) of a reagent being prepared in the laboratory **[D30:D31]**. Following equation 9,

$$wt \ (g) = M * mol \ wt * L$$

cell **[D32]** formula is:

=$D$30*$D$26*$D$31

The concentration **[D30]** is multiplied by the molar mass calculated in **[D26]** and the final volume needed **[D31]**. The user could additionally alter equation 9 to account for the purity of the substance as shown in equation 10. In this case, a new cell **[D32]** would be created for Purity of Substance (%) by inserting it between former lines 31 and 32 of both Figures 8.2 and 8.3. The grams of Solid Reagent Needed would then be calculated in a new cell (**[D33]**) using the equation:

= ($D$30*$D$26*$D$31*100)/$D$32

Examples: See problems 110b and 111b.

D. pH Titration Graph

The second application of spreadsheets extends the pH discussions in Chapter 4 to determining the pH of a solution by volumetric titrations. Titrimetric analysis involves the reaction of a known volume and concentration of a standard solution with a solution of the unknown concentration until the reaction is complete. For pH determinations, a standard solution of acid or base is added to a solution to determine conversely an unknown concentration of base or acid in the solution.

In Figure 4.1, a known volume and concentration of a weak base, Tris (25.00 mL of 0.1M Tris), is titrated with small volumes of 0.1M HCl to determine the concentration of the conjugate acid of Tris. At the inflection point of the curve, the concentrations of conjugate base and the conjugate acid are equal (equivalence point). Approximately 11 mL of 0.1N HCl were used to find the equivalence point.

In Figure 8.4, a 50.0 mL volume of 0.1 M HOAc solution is titrated against incremental volumes of Tris of unknown concentration to determine the equivalence point, and therefore the concentration of Tris. At each incremental volume of Tris added, the pH of the solution is measured using a calibrated pH meter. An indicator added to the solution of acid turned color at the 50.0 mL titrant volume, indicating a change from an acidic solution to a basic solution. The data is expressed in cells **[A2:B16]** in Figure 8.4. A graph of the data is illustrated on the spreadsheet, and is in itself a powerful tool within spreadsheets. All spreadsheet applications have graphing capabilities, but each is radically different in the steps needed to establish a graph. Refer to the software manual for your software application.

Notice the sharp rise around the 50.0 mL titrant volume, which corresponds to the change in indicator color observed above. An inflection point is a change in direction of the curve. As you increase the pH of the acetic acid solution, the curve traces counter-clockwise. At the sharp rise, the direction of the curve changes to a clockwise trace. The exact point of this change in direction is the inflection point, or equivalence point, and can be mathematically determined using a second derivative calculus equation. While the calculus is beyond the context of this material, there are defined algebraic equations that closely approximate the second derivative.

To find the equivalence point (EP), find the differences between successive pH readings, then calculate the differences between these values. When a sign change occurs, this is the inflection point. Thus, EP is defined as:

$$EP = -\{\{[(pH_1 - pH_2)*\Delta 1]/[\Delta 1 - \Delta 2]\} - pH_1\} \qquad (174)$$

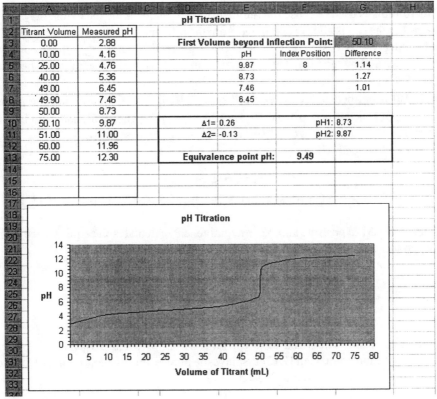

Figure 8.4.

where $\Delta2$ is the first negative difference value, $\Delta1$ = the value just before $\Delta2$, and pH_1 and pH_2 are the pH values on either side of the inflection point.

In Figure 8.4, the user enters the first volume beyond the estimated inflection point in cell **[G3]** (in this example 50.1 mL, because the color change of the indicator was observed at 50.0 mL). There will be some error checking to confirm this selection (explained below). Cell **[E5]** uses the formula:

$$= VLOOKUP(\$G\$3,A3:B16,2,FALSE)$$

to match the user-entered volume to the information in data table **[B2:C16]**. Cell **[G3]** is found in the table **[A3:B16]** ($\$G\$3,A3:B16$). The pH corresponding to the **[G3]** value from column 2 is returned in cell **[E5]**. FALSE forces an exact match of cell **[G3]** to the data options in column **[A-]** of the data. If an exact match is not found (or entered by the user), no value is returned.

Cell **[F5]** returns the row number in the data table of the pH looked up in **[E5]**,

$$= MATCH(E4,\$B\$2:\$B\$16)$$

Cell **[E6]** finds the pH value found above the pH in the data table from **[E5]** using the formula:

$$= OFFSET(\$B\$2,(\$F\$5-1))$$

This formula is used because the number of data points in the data table may vary, but the distance of the cell is always computed from the referenced column header $B$2. Under the pH heading **[E4]** is a list in reverse order of the pH readings starting with the corresponding volume entered by the user. The differences in successive pH readings can be found in cells **[G5:G7]**.

$\Delta 1$ **[E10]** is defined as the difference between **[G6]** and **[G7]**, and must be a positive number. $\Delta 2$ **[E11]** is defined as the difference between **[G5]** and **[G6]**, and must be a negative number. pH_1 is defined as cell **[E6]**, and pH_2 is defined as cell **[E5]**.

The formula in cell **[F13]**:

$$= IF(AND(\$E\$11<0,$$
$$\$E\$10>0),-(((((\$G\$10-\$G\$11)*\$E\$10)/(\$E\$10-\$E\$11))-\$G\$10),"")$$

checks that both $\Delta 1$ is positive and $\Delta 2$ is negative, and calculates the equivalence point pH result when these conditions are true, or displays a blank ("") if false.

The formula in cell **[E13]**:

$$= IF(AND(\$G\$11>\$F\$13,\$F\$13>\$G\$10),$$
$$\text{"Equivalence Point pH","Wrong Volume Chosen")}$$

verifies that the result calculated in **[F13]** falls within pH_1 and pH_2, the two pH values around the inflection point. Cell **[E13]** displays the phrase "Equivalence point pH" if conditions are true, or "Wrong Volume Chosen" if false.

Example: See problem 114.

Figure 8.5 illustrates a second example of the above spreadsheet, and should be referenced for problem 115.

Table M2 in Appendix M lists the cell contents for the spreadsheet template used in Figures 8.4 and 8.5.

Examples: See problems 113–115.

E. Dilution Preparation Worksheet

The third spreadsheet application is the creation of a Dilution Preparation Worksheet. This worksheet is based on a single equation, but uses error-checking logic to prevent erroneous results; much like in the two examples above, but in greater detail. This spreadsheet application is an example of a shared file or worksheet that many users could utilize in a laboratory setting.

In this example, the equation being solved is:

$$V_1C_1 = V_2C_2 \quad \text{or} \quad V_cC_c = V_dC_d \quad\quad (175)$$

but could be modified to solve any single-unknown, four-variable equation. For theoretical derivations, see Chapters 1, 3, and 4, specifically Charles' Law, Boyle's Law, Gay Lusaac's Law, the Combined Gas Law, and the buffer calculation examples.

The spreadsheet illustrated in Figure 8.6 contains four elements. The first element defines the equation in table form, i.e. the four variables are clearly labeled, each requiring a unit and numeric entry from the worksheet user **[B6:E7]**. The second element is an instruc-

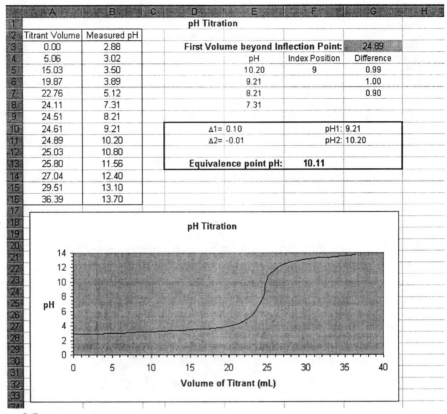

Figure 8.5.

tion list for the worksheet user to follow **[B9:B13]**. The third element is the solution to the defined equation **[B15:E15]**. The fourth element is an error checking array **[B18:E28]**.

Based on the instructions in the worksheet and the mathematical limitations of the equation, there are three requirements necessary to solve this single-unknown, four-variable equation.

1. Only one X can be solved for in the equation.
2. All data entry spots must be filled-in.
3. The units of both volume and concentration must match on both sides of the equation.

Requirement 1

The number of X's are counted in **[C20:C24]**. The cell formula for each variable:

$$= \text{IF}(\$B\$7="X",1,0)$$

assigns a value of 1 if an X is present in the cell, otherwise it assigns a value of 0. Cell **[C24]** sums the four assigned values for the number of X's.

Requirement 2

Requirement 2 follows Requirement 1, namely:

	A	B	C	D	E	F
1						
2			Dilution Preparation Worksheet			
3						
4			Concentrate		Dilute	
5		Volume, V_c	Conc., C_c	Volume, V_d	Conc., C_d	
6		mL	M	mL	M	
7		X	10	800	0.014	
8						
9		INSTRUCTIONS				
10		1. Place an 'X' in the cell of the variable with unknown value.				
11		2. Enter the units for volume and concentration.				
12		3. Complete the remaining information.				
13		4. The calculated result for 'X' is given below.				
14						
15			1.120 mL			
16						
17						
18			ERROR CHECKING			
19		Number of X's		Blank Units		
20		Vc	1	Vc	0	
21		Cc	0	Cc	0	
22		Vd	0	Vd	0	
23		Cd	0	Cd	0	
24		SUM	1	SUM	0	
25		Non-matching Units		Calculation	Units	
26		Volume	0	1	1	
27		Conc.	0	1	1	
28		SUM	0	2	2	
29						
30						

Figure 8.6.

$$= IF(\$B\$6="",1,0)$$

A value of 1 is assigned if the cell is blank, otherwise, a value of 0 is assigned. The sum of the four assigned values for the Blank Units is found in cell [E24]. If any of the equation variables are left blank the final result will not be calculated; thus it is unnecessary to check for blank cells.

Requirement 3

The volume and concentration units are checked for equality in cells [C26:C27]. For each unit, the cell formula:

$$= IF(\$B\$6=\$D\$6,0,1)$$

assigns a value of 0 if the units match, otherwise, assigns a value of 1. The sum of these two checks is in cell [C28].

"Units" in the Error Array determines if there are any blank cells or too many X's entered. If a blank cell is found, the total in cell [E24] will be greater than 0. If too many X's are entered, the total in cell [C24] will be greater than 1. If either of these conditions is found to be true, a value of 0 is assigned for each entry, otherwise a value of 1 is assigned. The sum in cell [E28] must total 2 for the equation requirements to be properly satisfied. (There are a number of ways to pick conditions and sum results to indicate specific situations; this is only one example. The user should use a system that is logical to him.)

"Calculation" in the Error Array assures that there are no blanks, and that the units match. The sum in cell **[D28]** must equal 2 for the requirements above to be properly satisfied.

It may appear that the items being checked under "Units" and "Calculation" above are reversed. This is intentional so that a result and unit will not be displayed if either does not meet the requirements above. The solution to the equation is defined in cells **[C15:C16]**. **[C15]** produces the numeric solution, using the following nested IF statements:

=IF($C$24>1,"Too many X's",IF($D$282,"",IF($B$7=
"X",($D8*$E8/$C8),IF($C$7="X",($D8*$E8/$B8),IF($D$7="X"
,($B8*$C8/$E8),IF($E$7="X",($B8*$C8/$D8),""))))))

If too many X's are present or **[D28]** does not equal 2, no value is calculated. The remaining formula is four nested IF statements, each checking in succession the location of the X to be solved for, finding the appropriate cell (**[B7]**, **[C7]**, **[D7]** or **[D8]**), then performing the simple equation. If by chance an X is not found (which should not be the case with the Error Array in place), a blank cell ("") is printed indicating that a solution was not found.

The unit of the unknown variable is found by locating the X, and assigning the unit directly above the value. The formula in **[D15]** first checks that unit requirements are met then finds the correct unit using nested IF statements.

=IF($E$282,"",(IF$B$7="X",$B$6,IF($B$7="X",$C$6,
IF($D$7="X",$D$6,IF($E$7"X",$E$6,""))))))

Table M3 in Appendix M lists the cell contents for the spreadsheet template used in Figure 8.6.

Examples: See problems 116–118.

F.   Routine Calculation Template

Perhaps the most practical spreadsheet application in laboratories is the creation of the routine calculation template. Calculations performed repeatedly can be encoded in a specific format, ensuring that relevant information and measured data are presented uniformly each time the calculation is needed. Routine calculation templates can be saved and recalled when needed, each time saving the information electronically under a different file name. This tool is often the cornerstone of a laboratory worksheet or report, and tends to be very straightforward and easy to prepare. Another use of these templates forms the basis of electronic data exchange with a LIMS, as demonstrated in section 8.3.

A spectrophotometric (or colorimetric) analysis method is a classic measurement technique based extensively on the Beer-Lambert law, as explained in section 5.2. This theoretical development can be extended to determine the concentration of an unknown substance, based on known or agreed-upon characteristics of a standard reference material. For an exact concentration of a substance named X, the equation:

$$A = E\ell C \tag{142}$$

holds true. E is a characteristic of substance X, and under the same experimental conditions (concentration unit and cuvette path length) remains constant. For an unknown concentra-

	A	B	C	D	E	F
1						
2			AMPROLIUM in FEEDS by COLORIMETRIC ANALYSIS			
3						
4	The Best Analytical Laboratory (TBAL)			TBAL Method:	12345	
5	1 Mole Way			AOAC Method:	961.24	
6	Anytown, Anystate 12345			Supervisor:	Dr. Supervisor	
7	Phone: (000) 123-4567					
8	FAX: (000) 123-4568			Analyst Name:	Jane Doe Chemist	
9				Date:	10/20/99	
10						
11			Amprolium Standard (USP Cat. No. 03400, Lot F)			
12		Purity (%):	100			
13		Mass (g):	0.0254			
14		Absorbance:	0.4736			
15						
16						
17	Sample Number	Sample Mass (g)	Extraction Volume (mL)	Absorbance	% Amprolium	
18	99-12345	15.0659	100.00	0.4126	0.01469	
19	Control 1	15.0927	100.00	0.4232	0.01504	
20	99-67890	5.0541	250.00	0.3498	0.09280	
21	00-00123	15.0709	100.00	0.5791	0.02061	
22	Control 2	15.3506	100.00	0.5818	0.02033	
23	00-00456	5.0541	250.00	0.5201	0.13798	
24	00-00789	5.0746	250.00	0.4892	0.12926	
25	00-01234	15.0416	100.00	0.4736	0.01689	
26						
27						
28	Signature:			Date:		
29						
30						

Figure 8.7.

tion of a solution of X, C_u, the absorbance, A_u, can be measured and equated to the absorbance (A_k) and concentration (C_k) of a standard as shown below.

$$C_k / A_k = 1 / E\ell = 1 / E\ell = C_u / A_u \qquad (176)$$

$$C_k A_u = C_u A_k \qquad (177)$$

Equation 177 forms the basis of the spreadsheet illustrated in Figure 8.7. Amprolium (an anti-coccidiosis additive in animal feeds) is determined spectrophotometrically relative to a single point calibration of an exact amount of a reference standard. (See Appendix A.3.) The samples are weighed, the drug extracted and the extraction purified. A chromophore is then added to both the samples and the standard, developing the chromophore-drug complex to a deep purple color. The measurement is made against a reagent blank in a dual-beam spectrophotometer. The %Amprolium is found using equation 178 below:

$$\%\text{Amprolium} = ((Wt_{std}*DF*Std_p*A_{sam})/((Wt_{sam}/Vol_{ext})*A_{std})*100) \qquad (178)$$

where Wt_{std} is the mass of standard, DF is the dilution factor (1.0E-5), Std_p is the purity of the standard material, A_{sam} is the absorbance of the sample, Wt_{sam} is the mass of the sample extracted, Vol_{ext} is the extraction volume, A_{std} is the absorbance of the standard, and 100 is the conversion factor. Equation 178 may not seem equivalent to Equation 177, but rearranging the variables in equation 178, and setting them equal to each other should make it more obvious.

$$\% \text{ Amprolium:} \quad \frac{A_{sam}}{A_{std}} = \frac{(Wt_{sam} / Vol_{ext})}{(Wt_{std} * Std_p * DF * 100)}$$

In Figure 8.7, equation 178 is represented in cells **[E18:E25]** as the following:

$$=(\$C\$13*0.001*\$C\$12*\$D\$18)/((\$B\$18/\$C\$18)/\$C\$14)$$

The user enters the laboratory and method information at the top of the template, the purity, mass and absorbance reading of the standard reference solution, and completes the sample number, sample mass, extraction volume, and sample absorbance reading. The % Amprolium is calculated. At this point the user can print the document, or forward it for review or electronically report the sample results to a LIMS.

Table M4 in Appendix M lists the cell contents for the spreadsheet template used in Figure 8.7.

Examples: See problems 119–120.

G. Calibration Regression Analysis

Section 8.2F described a spreadsheet application utilizing a single-point calibration. It is more likely the case in a laboratory that a series of calibration standards are utilized to determine the analytical response. This spreadsheet application is the most relevant of the examples in this chapter, applying to almost every area of chemical and biological analysis. It is, however, the most complicated, requiring a lengthy explanation. Multiple-point calibration is statistically more accurate than single-point calibration, thus, the mathematical theory is more rigorous. For a thorough treatment of the mathematics and statistics, see reference 21.

The mathematical equations are based on developing a line that best approximates the points on the calibration curve. The calibration curve consists of a measurand quantity, y, plotted against the known concentration, x, for a series of calibration standards. Regression Analysis is the method to determine the best-fit line encompassing these points. Specifically, the method of least-squares (linear least squares, LLS) is used to approximate a simple straight line. LLS is based on two assumptions. First a relationship between x and y exists and is described as

$$y = a + bx \qquad (179)$$

where y is the measured variable, a is the intercept (the value of y when x is zero), b is the slope of the line, and x is the known variable. Second, any deviation of the measurand quantity, y, is entirely due to random error; the known quantity is exactly known. These assumptions and equations form the basis of the theoretical development (reference 21).

In this spreadsheet application, five equations are used to determine the analytical response.

$$Sxx = \Sigma x^2 - (\Sigma x)^2/N \qquad (180)$$

$$Syy = \Sigma y^2 - (\Sigma y)^2/N \qquad (181)$$

$$Sxy = \Sigma xy - (\Sigma x \Sigma y)/N \qquad (182)$$

$$b = Sxy - Sxx \qquad (183)$$

$$a = ybar - b*xbar \qquad (184)$$

	A	B	C	D	E	F	G	H
1								
2		Microbiological Assay Response Regression Analysis						
3								
4	For all points on the curve:							
5	Standard Curve	Conc.	Zone Diameter	x2	y2	x*y		
6	Std. A	0.00733	16.99	5.37E-05	288.6601	0.124537		
7	Std. B	0.00917	17.80	8.41E-05	316.84	0.163226		
8	Std. C	0.01833	20.15	0.000336	406.0225	0.36935		
9	Std. D	0.02750	22.27	0.000756	495.9529	0.612425		
10	Std. E	0.03666	24.43	0.001344	596.8249	0.895604		
11		0.09899	101.64	0.002574	2104.3	2.165141		
12	N	5						
13	Sxx	0.00061421						
14	Syy	38.16248						
15	Sxy	0.15287228			Run Graphing Macro			
16	b=Sxy/Sxx	248.893226	(slope)					
17	a=y-bx	15.4004119	(y-intercept)					
18								

Figure 8.8a.

Σ is the symbol for sum. In equations 180 and 181, Sxx and Syy is found by summing the square of the x or y values $[\Sigma x^2$ or $\Sigma y^2]$, then subtracting the divisional product of the sum of x values squared divided by the number of pairs of data N $[(\Sigma x)^2/N$ or $(\Sigma y)^2/N]$. In equation 182, Sxy is found by summing the product of the x and y values of the data pair $[\Sigma xy]$, then subtracting the product of the individual x sums and y sums divided by the number of data pairs $[(\Sigma x\Sigma y)/N]$. These equations are illustrated in Figure 8.8a.

The five data pairs in the microbiological assay response are entered by the user in cells **[B6:C10]**. Cell **[B11]** sums the x values in the data pairs and **[C11]** sums the y values in the data pairs. Cells **[D6:D10]** determines the square of each individual x value (x^2), and sums all 5 values together in **[D11]**. The same for y is repeated in cells **[E6:E11]**. Likewise, the product of the data pairs x*y and their sum is found in **[F6:F11]**.

Sxx in cell **[B13]** calculates equation 180 as:

$$= \$D\$11 - (POWER(B11,2/B12)$$

The POWER function raises the value B11 to a power of 2, squaring the value. Syy in cell **[B14]** calculates equation 181 as:

$$= \$E\$11 - (POWER(C11,2)/B12)$$

Sxy in cell **[B15]** calculates equation 182 as:

$$= \$F\$11 - ((B11*C11)/B12)$$

The slope of the line, b, is found by dividing cell **[B15]** by **[B13]** as described in equation 183. The intercept in equation 184 is found by the formula in cell **[B17]**:

$$a = (C11/B12) - B16*(B11/B12)$$

where (C11/B12) is the average of the y values (sum/total pairs), B16 is the slope, and (B11/B12) is the average of the x values.

	A	B	C	D	E	F	G	H
19	Correlation Tests							
20					Is All_5_points RSQ above 0.9950?			YES
21	R-squared (all)	0.99702312	All_5_points		R-square value to be used:			0.99702312
22	R-squared (-A)	0.99947180	all_A		Lookup correct data array:			All_5_points
23	R-squared (-B)	0.99697435	all_B					
24	R-squared (-C)	0.99817629	all_C					
25	R-squared (-D)	0.99663017	all_D					
26	R-squared (-E)	0.99492554	all_E					
27		x	y				x	y
28	Regress. (all)	0.00733	16.99		Regress. (-C)		0.00733	16.99
29		0.00917	17.80				0.00917	17.80
30		0.01833	20.15				0.02750	22.27
31		0.02750	22.27				0.03666	24.43
32		0.03666	24.43					
33								
34	Regress. (-A)	0.00917	17.80		Regress. (-D)		0.00733	16.99
35		0.01833	20.15				0.00917	17.80
36		0.02750	22.27				0.01833	20.15
37		0.03666	24.43				0.03666	24.43
38								
39	Regress. (-B)	0.00733	16.99		Regress. (-E)		0.00733	16.99
40		0.01833	20.15				0.00917	17.80
41		0.02750	22.27				0.01833	20.15
42		0.03666	24.43				0.02750	22.27

Figure 8.8b.

At this point the user could graph the data points directly, as illustrated in Figure 8.8d. It is evident that a straight line reasonably approximates the calibration points. The concentration of an unknown substance can be determined from the calibration slope and intercept using equation 179.

Table M5 in Appendix M lists the cell contents for the spreadsheet template used in Figure 8.8.

In total, the information presented above is sufficient for Regression Analysis. Figure 8.8 has three additional sections: correlation tests (8.8b), secondary standard curve regression (8.8c), and graphing (8.8d).

	A	B	C	D	E	F	G	H
43	If (ALL POINTS) Curve RSQ not >=0.9950, use this abbreviated std. curve.; otherwise these cells are empty or non-functional.							
44								
45	Standard Curve	Conc.	Zone Diameter	x2	y2	x*y		
46				0	0	0		
47				0	0	0		
48				0	0	0		
49				0	0	0		
50		0	0	0	0	0		
51	N	4						
52	Sxx	0						
53	Syy	0						
54	Sxy	0						
55	b=Sxy/Sxx	#DIV/0!	(slope)					
56	a=y-bx	#DIV/0!	(y-intercept)					
57								

Figure 8.8c.

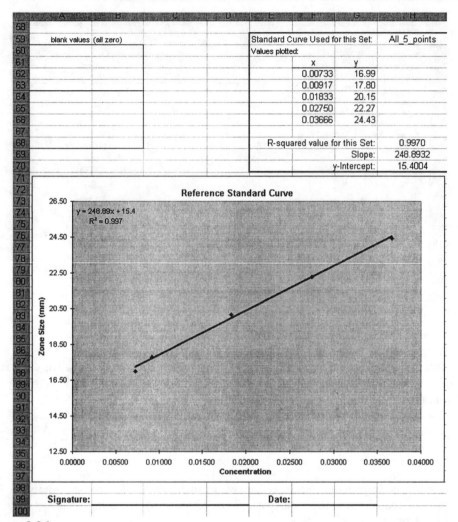

Figure 8.8d.

These sections are included because the laboratory from where this example originates defines in its Quality Assurance Plan minimum requirements data must meet in order to be reported. These requirements include:

1. Correlation of the LLS line (known as R^2) must be >0.995 (less than 0.5% error).

2. For methods utilizing biological organisms, if the R^2 value is <0.995, one data pair may be removed and the remaining 4 pairs reassessed. The reassessed data pairs must meet requirement 1 to report.

The data pairs (copied from the user entry in cells **[B15:C10]**) are reassembled in the correlation section, Figure 8.8b, rows **[-28]** through **[-42]**, removing one pair from each set. The R-squared values are determined for each set of data. The cell formula in cells **[B21:B26]** use the equation:

$$=RSQ(B28:B32,C28:C32)$$

where the cells **[B28:B32]** point to the x values and **[C28:C32]** point to the y values of the four data pairs from 'Regress-A' (first pair removed).

The box in the correlation section checks the regression of the line for all five data pairs. The formula in cell **[H20]** is:

$$=IF(\$B\$23>=0.995,"YES",'NO")$$

If $R^2>0.995$, YES is printed, and the R^2 value is listed in cell **[H21]**. If $R^2<0.995$, NO is printed in **[H20]**, and the maximum R^2 value of the abbreviated data pairs is determined by the formula in **[H21]**:

$$=IF(\$H\$20=YES,\$B\$21,MAX(B22:B26))$$

[H22] matches the selected R^2 values from **[H21]** to a name found in cells **[C21:C26]**. Cell **[H22]** formula is:

$$=VLOOKUP(\$H\$21,B21:C26,2,FALSE)$$

VLOOKUP looks up the value from cell **[$H$21]** in the first column in cells **[B21:C26]**, returning the cell contents in the second column of the row in which a match is found. The FALSE command requires an exact match between **[$H$21]** and the data in first column of the VLOOKUP.

If the selected data pair is one of the abbreviated data sets, it needs to be moved to the third section below to determine the slope and intercept of the LLS regression. Irrespective of the R^2 values, the data pair also needs to be moved to the graphing section. This process either must be performed by the user, or "programmed" into the spreadsheet by invoking a MACRO command.

The true power of spreadsheets is revealed in MACRO commands, where data can be manipulated based on logic functions. The MACRO command for Figure 8.8 is described in Appendix M, Table M6. This MACRO is essentially a fancy IF, THEN statement. It matches the data array name in cell **[H22]**, Figure 8.8b, then moves the data to the appropriate positions. Abbreviated data pairs are copied into cells **[B46:C49]** and **[F62:G66]**. Included in the logic is some error analysis, much like those included in previous examples. To utilize the MACRO, open the MACRO tools within the spreadsheet application and type in the code exactly as presented in Appendix M, Table M6. Secondly, create cell definitions within the spreadsheets as described in the comments section of the MACRO (first 14 lines). To define cell **[H22]** as "'data_array", highlight **[H22]** and select from the Menu Insert, Name. Type "data_array" (without the quotes) in the name position, and select Add. To verify that the cell is defined correctly, pick any other cell on the spreadsheet, then under the NAMEBOX in the FORMATTING TOOLBAR, pick "data_array" and the spreadsheet should highlight **[H22]**. Repeat this process for the remaining cell definitions. The MACRO should be run after the user enters the five original data pairs. The Button "RUN GRAPHING MACRO" (Figure 8.8a) calls the MACRO command, and performs the needed data manipulation. The MACRO command is also executed by pressing the alt+F8 keys.

The secondary regression (Figure 8.8c) performs LLS analysis of the abbreviated data pairs. The equations are exactly the same as described for the first section, with the noted exception that N is now 4.

	A	B	C	D	E	F	G	H
1								
2		Microbiological Assay Response Regression Analysis						
3								
4	For all points on the curve:							
5	Standard Curve	Conc.	Zone Diameter	x2	y2	x*y		
6	Std. A	0.00732	15.76	5.36E-05	248.3776	0.115363		
7	Std. B	0.00915	19.04	8.37E-05	362.5216	0.174216		
8	Std. C	0.01829	21.40	0.000335	457.96	0.391406		
9	Std. D	0.02744	23.21	0.000753	538.7041	0.636882		
10	Std. E	0.03659	25.03	0.001339	626.5009	0.915848		
11		0.09879	104.44	0.002564	2234.064	2.233715		
12	N	5						
13	Sxx	0.00061172						
14	Syy	52.52148						
15	Sxy	0.17018978				Run Graphing Macro		
16	b=Sxy/Sxx	278.216128	(slope)					
17	a=y-bx	15.3910057	(y-intercept)					
18								

Figure 8.9a.

The graphing section (Figure 8.8d) contains the summary box for the Microbiological Assay Response Regression Analysis. The Standard curve used is defined in cell **[H59]**, copied from cell **[H22]** above. Data pairs plotted in the graph appear in cells **[F62:G66]**. The R^2 value for the best-fit line appears in cell **[H68]**, copied from cell **[H21]**. The slope and intercept appear in cells **[H69]** and **[H70]**, respectively, using the formulas:

$$=IF(\$H\$20="YES", \$B\$16,\$B\$55)$$

$$=IF(\$H\$20="YES", \$B\$17,\$B\$56)$$

	A	B	C	D	E	F	G	H
19	Correlation Tests							
20					Is All_5_points RSQ above 0.9950?			NO
21	R-squared (all)	0.90152718	All_5_points		R-square value to be used:			0.995456396
22	R-squared (-A)	0.99545640	all_A		Lookup correct data array:			all_A
23	R-squared (-B)	0.93376672	all_B					
24	R-squared (-C)	0.92128787	all_C					
25	R-squared (-D)	0.88809053	all_D					
26	R-squared (-E)	0.86152529	all_E					
27		x	y				x	y
28	Regress. (all)	0.00732	15.76		Regress. (-C)		0.00732	15.76
29		0.00915	19.04				0.00915	19.04
30		0.01829	21.40				0.02744	23.21
31		0.02744	23.21				0.03659	25.03
32		0.03659	25.03					
33								
34	Regress. (-A)	0.00915	19.04		Regress. (-D)		0.00732	15.76
35		0.01829	21.40				0.00915	19.04
36		0.02744	23.21				0.01829	21.40
37		0.03659	25.03				0.03659	25.03
38								
39	Regress. (-B)	0.00732	15.76		Regress. (-E)		0.00732	15.76
40		0.01829	21.40				0.00915	19.04
41		0.02744	23.21				0.01829	21.40
42		0.03659	25.03				0.02744	23.21

Figure 8.9b.

	A	B	C	D	E	F	G	H
43	If (ALL POINTS) Curve RSQ not >=0.9950, use this abbreviated std. curve.; otherwise these cells are empty or non-functional.							
44								
45	Standard Curve	Conc.	Zone Diameter	x2	y2	x*y		
46		0.00915	19.04	8.37E-05	362.5216	0.174216		
47		0.01829	21.40	0.000335	457.96	0.391406		
48		0.02744	23.21	0.000753	538.7041	0.636882		
49		0.03659	25.03	0.001339	626.5009	0.915848		
50		0.09147	88.68	0.00251	1985.687	2.118352		
51	N	4						
52	Sxx	0.00041834						
53	Syy	19.651						
54	Sxy	0.0904622						
55	b=Sxy/Sxx	216.241852	(slope)					
56	a=y-bx	17.2250895	(y-intercept)					
57								

Figure 8.9c.

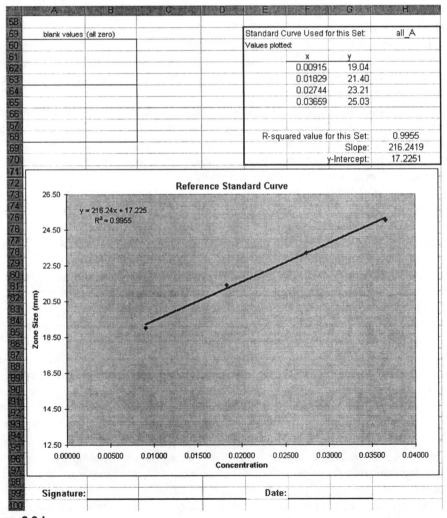

	A	B	C	D	E	F	G	H
58								
59	blank values	(all zero)			Standard Curve Used for this Set:		all_A	
60					Values plotted:			
61								
62					x	y		
63					0.00915	19.04		
64					0.01829	21.40		
65					0.02744	23.21		
66					0.03659	25.03		
67								
68					R-squared value for this Set:		0.9955	
69						Slope:	216.2419	
70						y-Intercept:	17.2251	

Reference Standard Curve

y = 216.24x + 17.225
R^2 = 0.9955

Zone Size (mm) vs Concentration

Signature:			Date:

Figure 8.9d.

If all five data pairs are used for the regression analysis, the slope and intercept is selected from the primary regression, top section, cells **[B16]** and **[B17]**. If an abbreviated data set is used for the regression, the slope and intercept is selected from the secondary regression section cells **[B55]** and **[B56]**.

Figure 8.9 illustrates a regression analysis in which an abbreviated data set is selected, and the secondary regression is utilized.

Examples: See problems 121–122.

8.3.   Reporting Results into LIMS

A Laboratory Information Management System (LIMS) is a network of computer hardware, software, laboratory equipment, and specialty instrumentation sharing the ability to exchange electronic information. The purpose of this network is sample management. LIMS can track all data input, perform simple calculations, create aliquot portions for special analysis, control instrument parameters and operations that generate analytical data, produce data reports, and audit all aspects of a chemical or biological process.

There are nearly 90 companies producing software products collectively called LIMS. The software vary significantly in scope and target science area. LIMS are designed for pharmaceutical, chemical, calibration, regulatory, global-corporation operations, web-based electronic access, and small laboratory applications, each having different user interaction and control mechanisms.

Generally, spreadsheets fit into LIMS as instruments, or file sources. This is in contrast to equipment and instruments that exchange data contemporaneously through data exchange protocols, such as RS-232 connections.

It is impossible to link each of these LIMS to the specific spreadsheet examples illustrated in this chapter. Furthermore, it would be improbable to link each LIMS to each spreadsheet application, such as Microsoft Excel, Lotus 1-2-3, or Corel Quattro or QuattroPro.

The computer industry has created two distinct file formats that are universally recognized by spreadsheet software developers. Each of the spreadsheet applications can save data into these file formats. Both file types contain only text that is based on ASCII characters and numbers, with only limited formatting such as font, type, and size. Graphics, lines, and figures are not allowed. The first format is known as a text file, commonly having the file extension, .txt. The second format is known as a delimited text file. A delimiter, such as a comma, space, tab, or another specific character separates the data in the file. There are two common file extensions for delimited spreadsheets: .csv, or comma separated variables, or .prn, or space separated variables.

Most LIMS will recognize the .txt, .prn, and .csv file types, and have methods to extract the information within these files. LIMS use control scripts that "parse," or selectively store pieces of data from the file. The file is navigated electronically, remembering specific items according to established rules within the LIMS. Data is then "mapped" internally to database records corresponding to the method and sample information from the .txt, .prn, or .csv file. It can be a complicated process.

The Amprolium example, section 8.2E, is the best spreadsheet example to demonstrate a simple electronic transfer of data. The example contains a list of sample numbers, and a final calculated answer expressing the final concentration of Amprolium found for each sample. If this spreadsheet template is created in Microsoft Excel, the saved file extension would be .xls. If the template is created in Lotus 1-2-3, the saved file extension would be

.wk4. Neither of these file types would be recognized by the LIMS. It is therefore necessary to save the analyte information as a text or delimited file.

Using Microsoft Excel as an example, save the spreadsheet in Figure 8.7 (problems 119 and 120) as another extension by selecting from the Menu File...Save As. Next to the prompt Save As Type, scroll through the options until the .txt or .csv extension is found. Click OK to save the file. Open this file through a simple word processing program, such as "Notepad". If the file was saved with a .csv extension, the textual version of the spreadsheet will appear as follows:

AMPROLIUM in FEEDS by COLORIMETRIC ANALYSIS,,,,

The Best Analytical Laboratory (TBAL),,,TBAL Method:,12345
1 Mole Way,,,AOAC Method:,961.24
"Anytown, Anystate 12345",,,Supervisor:,Dr. Supervisor
Phone: (000) 123-4567,,,,
FAX: (000) 123-4568,,,Analyst Name:,Jane Doe Chemist
,,,Date:,10/20/99
,"Amprolium Standard (USP Cat. No. 03400, Lot F)",,,
, Purity (%):,100 ,,
, Mass (g):,0.0254 ,,
,Absorbance:,0.4736 ,,
Sample Number,Sample Mass (g),Extraction Volume (mL),Absorbance,% Amprolium
99-12345,15.0659,100.00,0.4126,0.01469
Control 1,15.0927,100.00,0.4232,0.01504
99-67890,5.0541,250.00,0.3498,0.09280
00-00123,15.0709,100.00,0.5791,0.02061
Control 2,15.3506,100.00,0.5818,0.02033
00-00456,5.0541,250.00,0.5201,0.13798
00-00789,5.0746,250.00,0.4892,0.12926
00-01234,15.0416,100.00,0.4736,0.01689

Signature:,,,Date:,

A comma separates the contents of each cell within a row in the spreadsheet. The only information the LIMS should extract is the sample section. If spaces are substituted for Sample Mass (g), Extraction Volume (mL), and Absorbance, the data will immediately stand out.

Sample Number,1	,	,	,% Amprolium
99-12345,	,	,	,0.01469
Control 1,	,	,	,0.01504
99-67890,	,	,	,0.09280
00-00123,	,	,	,0.02061
Control 2,	,	,	,0.02033
00-00456,	,	,	,0.13798
00-00789,	,	,	,0.12926
00-01234,	,	,	,0.01689

A LIMS will then extract the sample number, match it to a database record, then attach the amount found to the sample record.

The advantages of LIMS and spreadsheets is clear. The majority of the work is done once by creating the template. The user opens the template, enters the sample information, and saves the file with a LIMS-compatible extension. The data generation is electronically traceable, and the calculation is never in doubt because the same validated template is used over and over. The data is not typed into another program, eliminating a possible transcription error. The chemist or biologist spends more time doing lab work, and less time duplicating work.

8.4.   Practical Examples

PROBLEM 109

Recreate the spreadsheet described in section 8.2B in your spreadsheet application. Refer to Appendix M for equivalent cell formulas, if necessary.

PROBLEM 110

(a) Calculate the molar masses of $CuCl_2 \cdot 2H_2O$ and Li_2SO_4 in problems 24 and 25 using the newly created spreadsheet in problem 109. (b) Calculate the mass in grams of solid reagent needed to prepare a 700.0 mL solution at 0.690M of each substance.

SOLUTION:

(b) 0.700 L * 170.49 g/mol * 0.69 M = 82.30 g $CuCl_2 \cdot 2H_2O$
     0.700 L * 109.98 g/mol * 0.69 M = 53.10 g Li_2SO_4

PROBLEM 111

(a) Using the Molecular Calculator Spreadsheet, what is the Molar Mass of lauryl sulfate (primary ingredient in shampoo), and the weight percent sodium in this molecule? (Molecular formula: $C_{12}H_{25}O_4SNa$.) (b) Calculate the mass in grams of solid reagent needed to prepare a 200.0 L mixing vat of aqueous lauryl sulfate at a concentration of exactly 2.1M.

SOLUTION:

(a) Molar mass: 288.4 g/mol
     Weight Percent Na: 7.97%
(b) 200 L * 288.4 g/mol * 2.1 M = 121,128 g $C_{12}H_{25}O_4SNa$
     Or   121.128 kg $C_{12}H_{25}O_4SNa$

PROBLEM 112

Using the Molecular Calculator Spreadsheet, what is the weight percent of carbonate, CO_3, in the molecule given in Figure 8.3?

SOLUTION:

Add another row at the bottom of the spreadsheet and copy the formulas into columns **[D-]** and **[E-]**. Calculate the molecular mass of CO_3, and enter this value in cell **[B30]**. Treat the carbonate as one unit, and complete the remaining entries in the Molecular Calculator.

Weight % CO_3:   35.70 %

PROBLEM 113

Recreate the spreadsheet described in section 8.2D in your spreadsheet application. Refer to Appendix M for equivalent cell formulas, if necessary.

PROBLEM 114

From the pH Titration Spreadsheet created in problem 113 and information in Section 4.2, what is the concentration of Tris buffer in the example found in Section 8.2D?

SOLUTION:

From the spreadsheet, at the equivalence point pH, the concentration of the conjugate acid and conjugate base are equal. The spreadsheet indicates that the actual volume of titrant at the equivalence point is between 50.0 mL and 50.1 mL. Using 50.1 mL as the volume of titrant,

$$(0.05010L \times 0.10 \text{ mol/L}) / 0.0500 \text{ L Tris} = 0.1002M \text{ Tris}$$

PROBLEM 115

From the information in Figure 8.5, what is the concentration of a solution of Aniline if 12.6 mL of this solution is titrated against 0.1M HCl acid?

SOLUTION:

From the spreadsheet, at the equivalence point pH, the concentration of the conjugate acid and conjugate base are equal. The spreadsheet indicates that the actual volume of titrant at the equivalence point is between 24.6 mL and 24.8 mL. Using 24.8 mL as the volume of titrant,

$$(0.02489L \times 0.10 \text{ mol/L}) / 0.0126 \text{ L Aniline} = 0.198 \text{ M Aniline}$$

PROBLEM 116

Recreate the spreadsheet described in section 8.2E in your spreadsheet application. Refer to Appendix M for equivalent cell formulas, if necessary.

PROBLEM 117

Using the spreadsheet created in problem 116, stepwise enter units and numeric variables for problem 88b. Note the contents in the result cells and the totals found in the Error Array. Purposely mismatch the units and add too many X's, all while observing the worksheet. Is it possible to obtain a result that is not valid?

SOLUTION

Doing this exercise is known as "validating" the worksheet. By forcing all possible conditions upon the logic, the spreadsheet can be tested for failures. If violations of the logic are found, the equations need to be altered to fix the violations. When all test conditions produce no failures in the logic, the equations in the spreadsheet should be compared to equations calculated by hand. If the spreadsheet results and the results calculated by hand are the same, then, and only then, is the spreadsheet considered validated for use within a laboratory setting.

PROBLEM 118

Change the variable descriptions in the worksheet created in problem 116 to calculate results for problems 56–61. Extend the number of variables on the spreadsheet to accommodate a single unknown, six-variable equation to solve problem 62.

PROBLEM 119

Recreate the spreadsheet described in section 8.2F in your spreadsheet application. Refer to Appendix M for equivalent cell formulas, if necessary.

PROBLEM 120

Alter the spreadsheet created in problem 119 to solve for the concentration of fat in several commercially available dog foods. The method information is given below, along with the observed measurements. Use the equation in Figure 8.6.

Standard material (98.4%) 0.01021g, Abs: 0.4162.

Samples are extracted with 100 mL 2:1 MeOH:H_2O. Standard and sample absorbance is measured on a single beam spectrophotometer.

Blank Abs: 0.0024
Sample 123: 10.1234g, Abs: 0.2695
Sample 456: 10.0428g, Abs: 0.4348
Sample 789: 10.0873g, Abs: 0.3874

SOLUTION:

Modify the % Drug column to subtract the blank measurement from both the standard and sample absorbance measurements.

Sample 123: 0.00641%
Sample 456: 0.01045%
Sample 789: 0.00927%

PROBLEM 121

Recreate the spreadsheet described in section 8.2G in your spreadsheet application. Refer to Appendix M for equivalent cell formulas.

PROBLEM 122

Using the spreadsheet created in problem 121, determine (a) the slope and y-intercept for the calibration data below, and (b) determine the final concentration of antibiotic if the observed measurand is 22.54 mmol.

Calibration Data	
Concentration (mg/mL)	**Measured (mmol)**
0.13758	13.21
0.22930	15.02
0.36689	17.69
0.45861	20.21
0.68791	25.19

SOLUTION:

Regression information from the spreadsheet is given below:

$R^2 = 0.9978$
a = 10.0096
b = 21.9497

Solving for x in equation 179, where y = 22.54mmol,

x = (y – a)/b
x = (22.54-10.0096)/21.9497
x = 0.5709 mg/mL antibiotic

References

1. Bloomfield, M.M. (1987) *Chemistry and the Living Organism*. John Wiley & Sons, New York.

2. Segal, B.G. (1989) *Chemistry, Experiment and Theory*, 2nd ed. John Wiley & Sons, New York.

3. Galister, H. (1991) *pH Measurement: Fundamentals, Methods, Applications and Instrumentation*. VCH Publishers, Inc., New York.

4. Georgiou, C.D., and Webster, D.A. (1987) *Biochemistry* **26**, 6521–6526.

5. Efiok, B.J.S., and Webster, D.A. (1990) *Biochem. Biophys. Res. Commun.* **173**, 370–375.

6. Zubay, G. (1988) *Biochemistry*, 2nd ed. Macmillan Publishing Co., New York. See pp 259–281 for a concise treatment of enzyme kinetics; and pp 236–238 for the application of spectrophotometry to monitor double- to single-stranded transitions of DNA.

7. Bashford, C. (1986). "An Introduction to Spectrophotometry and Fluorescence Spectrometry," in *Spectrophotometry & Spectrofluorimetry: A Practical Approach,* Bashford, C.L., and Harris, D.A., eds. IRL Press, Washington, D.C.

8. Cooper, T.G. (1977) *The Tools of Biochemistry*, John Wiley and Sons, New York, pp 59–62.

9. Lehninger, A.L. (1993) *Principles of Biochemistry*. Worth Publishers, Inc., New York. (See chapter on enzyme kinetics.)

10. Segel, I.H. (1976) *Biochemical Calculations*. John Wiley and Sons, New York. (A good reference for advanced and detailed treatment of biochemical calculations.)

11. Good, N.E., et al. (1966) *Biochemistry* **5**, 467–478; Good, N.E., and Izawa, S. (1968) *Hydrogen Ion Buffers. Methods in Enzymology* **24** (Pt. B), 53–68. (Sources for pK_a values.)

12. Friedlander, G., and Orr, W.C. (1951) *Phys. Rev.* **84**, 484; Laslett, L.J. (1949) *Phys. Rev.* **76**, 858. (Sources for isotope half-life.)

13. Bauer, G., et al. (1998) General Chemistry Handbook, Vol.II, 11th ed. Stipes Publishing Co., Champaign, IL, pp 50-51.

14. Bauer, G., et al. (1998) General Chemistry Handbook, Vol.I, 11th ed. Stipes Publishing Co., Champaign, IL, pp 99-100.

15. Hein, M. (1992) Foundations of College Chemistry, 4th Alternate ed. Brooks/Cole Publishing Co., Pacific Grove, California.

16. Brady, J.E. (1990) General Chemistry, Principles and Structure, 5th ed. John Wiley & Sons, New York.

17. Hill, J.W., and Petrucci, R.H. (1999) General Chemistry, An Integrated Approach, 2nd ed. Prentice Hall, New Jersey.

18. Zumdahl, S.S. (1997) Chemistry, 4th ed. D. C. Heath & Co., Lexington, MA.

19. Peters, E.I. (1990) Introduction to Chemical Principles, 5th ed. Harcourt Brace Jovanovich College Publishers, Fort Worth, TX.

20. Windows and Excel are either registered trademarks or trademarks of Microsoft Corporation in the United States. Quattro and Quattro Pro are registered trademarks of Corel Corporation in the United States. Lotus 1-2-3 is a registered trademark of Lotus Development Corporation in the United States.

21. Skoog, D.A., West, D.M., Holler, F.J. (1988) *Fundamentals of Analytical Chemistry*, 5th ed. Saunders College Publishing, New York, 39–42.

Appendix A

Supplements to Text Materials

1. Derivation of the Ionization Constant of H_2O

Pure H_2O ionizes as follows:

$$H_2O \rightleftharpoons H^+ + OH^-$$

and the equilibrium constant is

$$K_{eq} = \frac{[H^+][OH^-]}{[H_2O]}$$

Rearrangement yields

$$K_{eq}[H_2O] = [H^+][OH^-]$$

The concentration of H_2O in aqueous solution is practically constant (about 55.6 M) because at $25°C$, only about 10^{-7} M dissociates. Therefore, the term, $K_{eq}[H_2O]$ is a constant and is defined as K_w, the dissociation constant of H_2O.

$$\therefore K_w = [H^+][OH^-]$$

2. Derivation of the Henderson-Hasselbach Equation

A weak acid ionizes in H_2O as follows:

$$HA + H_2O \rightleftharpoons H_3O^+ + A^-$$

The ionization constant (K_i) is:

$$K_i = \frac{[H_3O^+][A^-]}{[H_2O][HA]}$$

Multiplying both sides of the equation by $[H_2O]$ yields:

$$K_i[H_2O] = \frac{[H_3O^+][A^-]}{[HA]}$$

[H$_2$O] is large (55.6 M) and essentially constant (in pure H$_2$O at 25°C, only 10^{-7} M is ionized). Therefore, K_i[H$_2$O] is a constant defined as K_a, the acid dissociation constant. Also, [H$_3$O$^+$] = [H$^+$]. Substituting these into the above equation yields:

$$K_a = \frac{[H^+][A^-]}{[HA]}$$

Taking the logarithm of the above equation:

$$\log K_a = \log \frac{[H^+][A^-]}{[HA]} = \log ([H^+][A^-]) - \log [HA] = \log [H^+] + \log [A^-] - \log [HA]$$

$$-\log [H^+] = -\log K_a + \log [A^-] - \log [HA] \quad -\log [H^+] = -\log K_a + \log \frac{[A^-]}{[HA]}$$

By definition, pH = $-\log$ [H$^+$] and pK_a = $-\log K_a$. Substituting these into the above equation yields:

$$pH = pK_a + \log \frac{[A^-]}{[HA]}$$

This is the Henderson-Hasselbach equation, written for a weak acid. For a weak base (R-NH$_2$), the equation becomes:

$$pH = pK_a + \log \frac{[R - NH_2]}{[R - NH_3^+]}$$

3.   Derivation of the Beer-Lambert Equation

Lambert's law states that when monochromatic incident light of intensity I_0 passes through a solution of a light-absorbing substance, the intensity (I) of the transmitted light decreases as an exponential function of the path length ℓ (Figure A.3-1).

From the above definition, $I = I_0 \, 10^{-k\ell}$,

$$\therefore \frac{I}{I_0} = 10^{-k\ell}$$

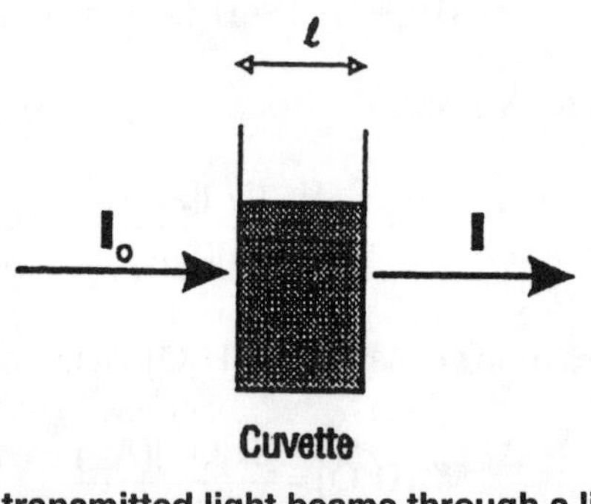

Cuvette

Figure A.3-1.   Incident and transmitted light beams through a light-absorbing solution.

Similarly, Beer's law states that I decreases as an exponential function of the concentration (C) of the light-absorbing substance. Thus,

$$I = I_0\, 10^{-k'C},$$

$$\therefore \frac{I}{I_0} = 10^{-k'C}$$

Combining the Lambert and Beer equations,

$$\frac{I}{I_0} = 10^{-E\ell C}$$

where E is a combined constant for k and k'.

Upon taking the logarithm,

$$\log \frac{I}{I_0} = \log 10^{-E\ell C} = -E\ell C - \log = \frac{I}{I_0}\, E\ell C, \text{ or } \log \frac{I_0}{I} = E\ell C$$

The term, $\log (I_0/I)$, is defined as absorbance (A),

$$\therefore A = E\ell C$$

This is the Beer-Lambert equation. For transmittance (T),

$$T = \frac{I}{I_0}$$

Also,

$$\%T = \frac{I}{I_0} \times 100\%$$

4.   Spectrophotometers and Absorbance Measurement

A spectrophotometer is an instrument used to measure the optical absorbance or transmittance of a substance. It measures the incident and emergent light beams and calculates the absorbance or transmittance using these measured values (see Appendix A.3). The instrument can be classified as single beam or dual beam.

A.   Single-Beam Spectrophotometers

These machines use a single light path for measurements. Their five essential components (Figure A.4-1) are as follows:

(1) Light source: A tungsten or Xenon arc lamp is commonly used for measurements in the visible region (400–900 nm), and a hydrogen lamp is used for measurements in the UV region (200–400 nm). UV-visible spectrophotometers have both types of lamp, and selection is either automatic or manual. The major requirement for these lamps is that they have stable and high-intensity output.

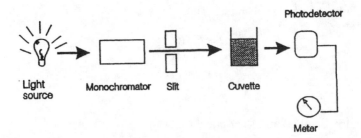

Figure A.4-1. Components of single-beam spectrophotometers. See text for explanation.

(2) Monochromator: A light beam from the lamp passes through a monochromator assembly where it is separated into a continuous spectrum of wavelengths by a prism or a diffraction grating. By adjusting the wavelength selector on the machine, a slice of the light spectrum centering about the wavelength of choice is isolated and passed through a slit (exit) of the monochromator assembly. (The monochromator does not select light of a single wavelength but a small spectrum of light of the same color having a maximum intensity at the chosen wavelength. In older machines, the monochromator is simply a filter that transmits light of the same color.)

(3) Slit: The incident beam exiting the monochromator passes through a slit that controls the wavelength purity as well as the intensity of the beam. The slit width may be fixed or adjustable (most simple spectrophotometers have fixed slit width). Decreasing the slit width increases the wavelength purity and, as such, the resolution or specificity of the measurement. However, this leads to a decrease in the intensity of the incident light and results in a decreased sensitivity. The reverse occurs when the slit width is increased. Choosing the right slit width, therefore, involves striking a balance between the desired resolution and sensitivity.

(4) Sample cuvette: From the slit, the incident beam passes through a sample cuvette positioned inside a sample housing. The housing is covered to prevent external light from reaching the photodetector and interfering with measurements. When more than one cuvette is used to measure a set of samples, the cuvettes should be optically identical or balanced so that errors arising from differences in their path length and light transmission properties cancel out. Quartz or silica cuvettes are used when measuring absorbance in the UV region because they do not absorb light significantly either in the UV or the visible region. Glass cuvettes absorb strongly in the UV region and are therefore unsuitable, except for measurements in the visible region where they do not absorb significantly.

(5) Light detector/display devices: Light transmitted through the sample is measured and amplified by a photosensitive detector (photo multiplier, photocell/amplifier pair or photo-detector array), and the signal is converted to absorbance or transmittance and displayed on a meter or a chart recorder.

B. Dual-Beam Spectrophotometers

Dual-beam spectrophotometers differ from single-beam machines in that they use two light paths for measurements: The incident light leaving the monochromator is split into a reference beam and a sample beam. The reference beam is passed through a cuvette containing a blank or reference solution, and the sample beam is passed through a cuvette containing a sample. Light emerging from each cuvette is detected by a photodetector, and the machine automatically subtracts the reference signal from the sample signal and uses the net value to calculate the sample's absorbance, transmittance, or concentration. The dual-beam technique compensates for errors due to fluctuations in the intensity of the lamp's output.

Some-dual beam spectrophotometers are simple machines, but others are sophisticated: They are capable of automatically scanning the UV and visible wavelength regions and simultaneously recording the absorbance, transmittance, or concentration of a sample. Additionally, they have optional components and internal microprocessors that facilitate fine-tuning the sensitivity and specificity of measurements, as well as storing and analyzing scanned spectra.

C. General Procedure For Absorbance Measurement

(1) Single-Beam Simple Spectrophotometers

(a) Turn on the machine and let it warm up for 15–20 minutes. If a UV wavelength will be used, also turn on the hydrogen lamp.

(b) Select the wavelength using the wavelength selector.

(c) Select the appropriate cuvette: quartz or silica for UV/visible, or glass for visible wavelengths.

(d) Calibrate the absorbance/transmittance scale as follows:

- With no cuvette in position, set the absorbance to ∞, or the transmittance to zero: When a cuvette is not inserted, the incident light is prevented from reaching the photodetector. The signal across the photodetector represents the maximum absorbance or zero transmittance.

- Fill the cuvette with an adequate volume of a blank solution and position it in the sample housing. With the housing door closed, set the absorbance to zero using the zero adjust, or the transmittance to 100%. The blank solution should best approximate the test solution, but lacks the assay's test color.

(e) Discard the blank solution, rinse the cuvette and fill it with the sample or test solution. Insert the cuvette as before and read the absorbance or % transmittance of the sample.

(2) Dual-Beam Spectrophotometer

Specific instructions for using these machines are best described in their operation manuals. The basics are similar to that described above except that:

(a) The blank solution should be in both the reference and the sample positions before setting the zero absorbance.

(b) When measuring the absorbance for samples, the blank should always be in its position.

(c) Some machines set the maximum absorbance automatically.

5. Derivation of the Relationship Between [S], [S₀], and k for a First-Order Rate Reaction

Equation 149 showed that when $[S]$ is limiting in an enzyme assay, the velocity (v) is proportional to the substrate concentration ($[S]$):

$$v = k[S]$$

k being the proportionality constant. The velocity is also defined as the decrease in $[S]$ per unit change in time:

$$v = -\frac{d[S]}{dt}$$

Combining the two equations,

$$v = k[S] = -\frac{d[S]}{dt}$$

or

$$-\frac{d[S]}{dt} = kdt$$

This equation is then integrated from $[S_0]$ to $[S]$ for the left side, and from 0 to t for the right side; $[S_0]$ and $[S]$ being the substrate concentrations at time $t = 0$ and a later time t, respectively.

$$-\int_{S(o)}^{S(t)} \frac{d[S]}{[S]} = k\int_{o}^{t} dt$$

$$-(\ln [S] - \ln [S_0]) = k\,(t - 0)$$

$$\ln \frac{[S_0]}{[S]} = kt$$

Taking the antilog,

$$\frac{[S_0]}{[S]} = e^{kt}$$

$$\therefore [S_0] = [S]e^{kt} \text{ or } [S] = [S_0]e^{-kt}$$

The change in substrate concentration ($\Delta[S]$) within the time interval t_1 to t_2 can be related to $[S_0]$. Starting from the last equation,

$$\Delta[S] = [S_{t_1}] - [S_{t_2}] = [S_0]e^{-k_{t_1}} - [S0]e^{-k_{t_2}}$$

i.e., $$\Delta[S] = [S_0](e^{-k_{t_1}} - e^{-k_{t_2}})$$

or

$$[S_0] = \frac{\Delta[S]}{(e^{-k_{t_1}} - e^{-k_{t_2}})}$$

6. Liquid Scintillation and Geiger-Müller Techniques for Radioactivity Measurements

The general principle underlying these techniques for radioactivity measurement is as follows: When a ß particle collides with an atom, the atom absorbs the particle's energy and becomes either ionized or excited to a higher energy state. Liquid scintillation counting is based on detecting the fluorescence that accompanies the return of the excited atom to ground state, whereas the Geiger-Müller technique is based on detecting the ionized atom.

A. Geiger-Müller Technique

This technique counts mostly ß particles. In general, a ß particle enters a gas-filled tube (Geiger tube) in which an electric field has been applied. Upon collision with an atom of the gas, the particle's energy is absorbed by the atom which then loses an electron and produces an ion pair — an electron that is attracted to the anode, and a positively charged ion of the gas that is attracted to the cathode. The electrodes detect and record these ions as electrical pulses, representing counts.

B. Liquid Scintillation Technique

This is used to count both ß and γ particles, but the detector used for each type of particle differs.

(1) ß Counting: The overall strategy is to set up an assay medium in which its components will efficiently transfer the energy of a ß particle to fluorescence molecules that emit light of a wavelength suitable for detection by a photodetector. In practice, the radioactive sample is mixed with a liquid scintillation "cocktail" that contains an excitable solvent (usually toluene) and one or more fluorescent compounds short-named "fluors." A ß particle emitted by a radioactive sample collides with and transfers some or all of its energy to a solvent molecule which becomes excited. The excited solvent molecule either transfers its energy to another solvent molecule, or phosphoresces and emits the excitation energy as light having a wavelength that is usually too short for detection by the instrument's phototube. This photon is absorbed by a primary fluor (F_1), which becomes excited, fluoresces, and emits a photon of longer wavelength. If its wavelength is sufficiently long, the photon is detected and counted by the phototube; otherwise, it is reabsorbed by a secondary fluor, which subsequently fluoresces and emits a photon with a longer wavelength for a subsequent detection. If the wavelength is still too short, additional fluors with suitable characteristics are added to the cocktail. A summary of the net reaction is as follows:

$$\text{solvent} + \text{ß}^- \rightarrow \text{ß (lower energy)} + \text{solvent*}$$

$$\text{solvent*} + F_1 \rightarrow \text{solvent} + F_1\text{*}$$

$$F_1\text{*} \rightarrow F_1 + (h\nu)_1$$

$$\text{or } F_1\text{*} + F_2 \rightarrow F_1 + F_2\text{*}$$

$$F_2^* \rightarrow F_2 + (h\nu)_2$$

where $(h\nu)$ is the photon emitted by the fluorescent compounds, which is detected by the phototube and reported as counts per minute.

(2) γ Counting: The liquid scintillation fluid described above for ß particles is unsuitable for counting γ photons. This is because a γ photon does not have mass; hence, it penetrates matter deeply and, therefore, requires a medium denser than a liquid to be absorbed efficiently. A fluorescent NaI crystal cell is the medium commonly used. The sample (usually a solid in a vial) is placed in the NaI cell. The γ photon leaving the sample penetrates and interacts with the NaI crystal producing ß particles which excite adjacent portions of the crystal and cause fluorescence. Light photons from the fluorescence are detected and counted by the phototube.

Appendix B

Dealing with Numbers

1. Handling Exact and Experimental Numbers

Data used for laboratory calculations consist of exact and/or measured numbers. Exact numbers are those whose values are exactly known. They include numbers that result from fundamental definition of quantities or from counting objects. For example, 1 L = 1000 mL; the 1000 is an exact number because its exact value is known. Exact numbers have an infinite number of **significant digits** or **significant figures**. If this were to be indicated numerically, the 1000 above would be written as 1,000.00000 . . . infinity. For convenience, the zeros to the right of the decimal point are omitted, but their presence must be assumed when deciding the number of significant figures in a value calculated from data containing both exact and measured values.

In contrast, numbers that are experimentally measured are not exact because of small errors or uncertainties associated with laboratory measurements. All laboratory instruments and techniques have limited accuracy, which in turn imparts small errors or uncertainty to experimental values measured with them. Scientists indicate these uncertainties by using an appropriate number of significant digits or significant figures in the measured values **(a significant digit is one whose value is reasonably reliable)**. This is done by recording all digits that are known with certainty, and then adding an extra digit that has an uncertainty. To illustrate, if the light absorption of a sample is being measured using a spectrophotometer, and the meter indicates the reading shown in Figure B.1-1, then the first two digits of the result (0.52) are certain because the indicator points between the calibration marks 0.52 and 0.53. The third digit is uncertain because the space between the two numbers is not clearly marked. Its estimated value is 6, and unless specified otherwise, an estimated digit is assumed to have a range of ±1. Therefore, the result is recorded as 0.526, but its exact value lies between 0.525 and 0.527. Estimating the result to four decimal places such as 0.5258 does not improve its accuracy because the instrument is limited to only three decimal places.

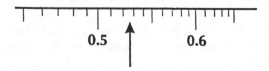

Figure B-1.1. Recording experimental data.

2.   Determining Significant Figures

In any number, the digits 1 through 9 are all significant. Thus the numbers 561 and 0.56123 have three and five significant figures, respectively.

The digit 0 may or may not be significant, depending on where it appears in a number.

(a) **If the number is less than 1, all zeros to the right of the first nonzero digit are significant, but zeros to the left are not.** For example, the number 0.037700 has five significant figures because the two zeros that precede the first nonzero digit 3, are not significant, whereas the two zeros to the right are.

(b) **If the number is greater than 1, and there are digits on both sides of the decimal point, then all zeros are significant.** Example, in 5.0003 and 50.300, all the zeros are significant, hence, each has five significant figures.

(c) **If the number is greater than 1 and there are no digits to the right of the decimal point, then zeros that are located between two nonzero digits are significant. Zeros to the right of the last nonzero digit are not significant unless indicated otherwise.** For example, 70,029 has five significant figures because the two zeros between the digits 7 and 2 are significant. Each of the numbers 729, and 72,900 has three significant figures because the zeros in 72,900 are assumed not significant. If these zeros are significant, the number is expressed in exponential form, 7.2900×10^4, to indicate five significant figures.

3.   Rounding Off Numbers

When experimental data or calculated results have more digits than are needed to indicate reliability, they are **rounded off** by eliminating the excess digits. The rules for rounding off are summarized below.

(a) If the first digit dropped is greater than a 5, increase the last retained digit by 1. For example, 3.3772 becomes 3.38 when rounded to two decimal places or three significant figures.

(b) If the first digit dropped is less than a 5, do not change the last retained digit. Using the above example, 3.3772 becomes 3.377 when rounded off to three decimal places or four significant figures.

(c) If the last digit dropped is a 5, do as follows: If the 5 is followed by digits greater than zero, increase the last retained digit by 1. For example, 3.88512 becomes 3.89 when rounded to two decimal places.

If the 5 is followed by a zero or no other digit, increase the last retained digit by 1 if that digit is odd, but leave it unchanged if it is even. For example, 3.885 or 3.88501 become 3.88 when rounded to three digits. In contrast, 3.875 or 3.8750 becomes 3.88 when rounded to three digits.

4.   Significant Digits and Rounding in Calculated Values

Results calculated from experimentally derived data should retain an appropriate number of significant figures or decimal places to indicate the reliability of the calculated values. Here are the rules:

(a) Decide how many significant figures or decimal places the final answer should have [see rules (c), (d), and (e), below].

(b) If a result is to be used for further calculations, retain one or more significant digits than is necessary. If the result is final, round it to the number of significant figures estimated in step (a).

(c) If the calculations involve division and/or multiplication only, report the final result to as many significant figures as there are in the term with the least number of significant figures. For example,

$$\frac{37.0}{36.51 \times 0.08403} = 12.06 = 12.1$$

The 12.1 has been rounded to the same number of significant digits as 37.0; the factor with the least number of significant figures.

(d) If the calculations involve addition and/or subtraction, report the final result to as many decimal places as there are in the term with the least number of decimal places. For example,

$$12.046 + 9.2 - 0.345 = 20.901 = 20.9$$

The answer has been rounded to 20.9 because the term 9.2 has the least number of decimal places, which is one.

(e) If the calculations involve division/multiplication and/or addition/subtraction, follow the above rules, but perform the addition/subtraction first. In the example below, the addition operation in the numerator is performed first. Because its answer is to be used for a further calculation, it retains one more significant figure than would be required if it were the final answer. The division is performed last, and the final answer, 7.7, has been rounded to one decimal place in accordance with the above rule.

$$\frac{0.5 + 0.075}{0.075} = \frac{0.575}{0.075} = 7.6667 = 7.7$$

5.   Exponential (Scientific) Notation

Larger and smaller units for the same measurement sometimes differ by multiples of 10 (especially in the metric system), such as 10 ($*10^1$), 100 ($*10^2$), 1000 ($*10^3$), 10000 ($*10^4$), or 0.001 ($*10^{-3}$), 0.00001 ($*10^{-5}$), etc. Very large or small numbers are often used in chemistry. It is often convenient to express these numbers using the power of 10 or as exponentials (* shown here). For example, the mass of a helium atom is 0.0000000000000664 gram. In one liter of helium at $0°C$ and 1 atmosphere of pressure there are 26,880,000,000,000,000,000,000 helium atoms. These numbers look awkward as written, and to avoid this awkward expression, exponential or scientific notation is used to express them.

To write a number in scientific notation, move the decimal point in the original number so that it is located after the first nonzero digit. This new number is multiplied by 10 raised to the proper power (exponent). The power of 10 is equal to the number of places that the decimal point has been moved. If the decimal was moved to the left, the power of 10 will be a positive number. If the decimal was moved to the right, the power of 10 will be a negative number.

Examples:

(a) Express 5293 in exponential (or scientific) notation.

Solution: Looking at the number 5293. The decimal or period (.) is on the extreme right. We now have to place this decimal between the 5 and the 2. To do this the decimal will be moved three places to the left, so the power of 10 will be 3, and the number 5.293 is multiplied by 10^3. The correct scientific notation is 5.293×10^3.

(b) Express 4,500,000,000 in scientific notation with two significant figures.

Solution: 4,500,000,000. Place the decimal between the 4 and the 5. Since the decimal was moved 9 places to the left, the power of 10 will be 9, and the number 4.5 is multiplied by 10^9.

4.5×10^9 (Correct scientific notation).

(c) Express 0.000123 in scientific notation.

Solution: 0.000123. Place the decimal between the 1 and the 2. Since the decimal was moved 4 places to the right, the power of 10 will be -4, and the number 1.23 is multiplied by 10^{-4}.

1.23×10^{-4} (Correct scientific notation).

6.   Multiplying and Dividing Numbers in Exponential Notation

When two exponential with the same base are multiplied, the product is the same base raised to a power equal to the sum of the exponents. When exponentials are divided, the denominator exponent is subtracted from the numerator exponent:

$$10^a \times 10^b = 10^{(a+b)}$$

$$10^a \div 10^b = \frac{10^a}{10^b} = 10^{(a-b)}$$

In multiplying or dividing numbers in the exponential notation, we rearrange the factors. All coefficients are placed in one group and all exponents are placed in another group. The two groups are evaluated separately. The decimal result and the exponential result are then combined. For example,

$$(3.95 \times 10^5)(5.23 \times 10^{-8})$$

First remove the parenthesis.

$$= 3.95 \times 10^5 \times 5.23 \times 10^{-8}$$

Then regroup;

$$= (3.95 \times 5.23)(10^5 \times 10^{-8})$$

Next, evaluate the groups separately.

$$= 20.7 \times 10^{-3}$$

Finally, relocate decimal point.

$$= 2.07 \times 10^{-2}$$

Similarly,

$$(3.95 \times 10^4) \div (5.23 \times 10^{-8}) = \frac{3.95}{5.23} \times \frac{10^4}{10^{-8}}$$

$$= 0.755 \times 10^{12} = 7.55 \times 10^{11}$$

Appendix C

International System of Units (SI)

With a few exceptions, most of the units used in this book are SI (Systeme Internationale). It is a system of units adopted in 1960 by an international organization to bring world-wide uniformity to scientific measurements. The basis for some of the units are shown in Table C1, and the prefixes used to indicate their fractions and multiples are shown in Table C2. For example, the prefixes for 1×10^{-3} and 1×10^{-6} of a unit are milli-(m) and micro-(μ), respectively. For the mass unit gram (g), 1 milligram (mg) $= 1 \times 10^{-3}$ g and 1 μg $= 1 \times 10^{-6}$ g. Similarly, for the amount unit mole (mol), 1 mmol $= 1 \times 10^{-3}$ mol, and 1 μmol $= 1 \times 10^{-6}$ mol. See Quick Reference to Units (page xiv) and Tables C3 through C6 below for more examples.

Instructions for Tables C3–C6

Definitions are: Source unit; the unit to be converted. Target unit; the unit into which the source unit is to be converted. To convert from one unit to another, select the **source unit** from the **boldfaced** set (leftmost column). Move in a straight line across the corresponding row until you reach a factor directly under the *italicized target unit* (topmost row). Multiply your data by this factor. For example, to convert 0.5 **mg** to μg, the **mg** is selected from the boldfaced set in the leftmost column of units in Table C3. Moving straight across the corresponding row, the conversion factor 10^3 is directly under the *italicized* μg in the top row of units. This means that:

$$1 \text{ mg} = 10^3 \text{ }\mu\text{g}; \therefore 0.5 \text{ mg} = (0.5 \times 10^3) = 500 \text{ }\mu\text{g}.$$

If the source unit is selected from the *italicized* set instead, follow the above instructions except that the target unit is selected from the boldfaced set, and the data is divided by the factor: Using the above example, the factor obtained is 10^{-3}.

$$\therefore 0.5 \text{ mg} = 0.5 \div 10^{-3} = 500 \text{ }\mu\text{g}.$$

Table C1. Selected SI Units

Quantity	Unit	Symbol
Mass	kilogram	kg
Volume	liter	L or l
Amount of substance	mole	mol
Length	meter	m
Time	second	s
Electric current	ampere	A
Temperature	Kelvin	K

Table C2.   Prefixes for SI Units

Prefix	Symbol	Multiples and submultiples
Exa	E	10^{18}
Peta	P	10^{15}
Tera	T	10^{12}
Giga	G	10^{9}
Mega	M	10^{6}
Kilo	k	10^{3}
deci	d	10^{-1}
centi	c	10^{-2}
milli	m	10^{-3}
micro	μ	10^{-6}
nano	n	10^{-9}
pico	p	10^{-12}
femto	f	10^{-15}
atto	a	10^{-18}

Table C3.   Conversion Factors for Gram Units

Source	Target						
	kg	*g*	*mg*	*μg*	*ng*	*pg*	*fg*
kg	$\mathbf{10^0}$	10^{3}	10^{6}	10^{9}	10^{12}	10^{15}	10^{18}
g	10^{-3}	$\mathbf{10^0}$	10^{3}	10^{6}	10^{9}	10^{12}	10^{15}
mg	10^{-6}	10^{-3}	$\mathbf{10^0}$	10^{3}	10^{6}	10^{9}	10^{12}
μg	10^{-9}	10^{-6}	10^{-3}	$\mathbf{10^0}$	10^{3}	10^{6}	10^{9}
ng	10^{-12}	10^{-9}	10^{-6}	10^{-3}	$\mathbf{10^0}$	10^{3}	10^{6}
pg	10^{-15}	10^{-12}	10^{-9}	10^{-6}	10^{-3}	$\mathbf{10^0}$	10^{3}
fg	10^{-18}	10^{-15}	10^{-12}	10^{-9}	10^{-6}	10^{-3}	$\mathbf{10^0}$

Table C4.   Conversion Factors for Liter Units

Source	Target					
	kL	*L*	*dL*	*mL*	*μL*	*nL*
kL	$\mathbf{10^0}$	10^{3}	10^{4}	10^{6}	10^{9}	10^{12}
L	10^{-3}	$\mathbf{10^0}$	10^{1}	10^{3}	10^{6}	10^{9}
dL	10^{-4}	10^{-1}	$\mathbf{10^0}$	10^{2}	10^{5}	10^{8}
mL	10^{-6}	10^{-3}	10^{-2}	$\mathbf{10^0}$	10^{3}	10^{6}
μL	10^{-9}	10^{-6}	10^{-5}	10^{-3}	$\mathbf{10^0}$	10^{3}
nL	10^{-12}	10^{-9}	10^{-8}	10^{-6}	10^{-3}	$\mathbf{10^0}$

Table C5.   Conversion Factors for Meter Units[a]

Source	Target						
	km	*m*	*cm*	*mm*	*μm*	*nm*	*Å*
km	**10^0**	10^3	10^5	10^6	10^9	10^{12}	10^{13}
m	10^{-3}	**10^0**	10^2	10^3	10^6	10^9	10^{10}
cm	10^{-5}	10^{-2}	**10^0**	10^1	10^4	10^7	10^8
mm	10^{-6}	10^{-3}	10^{-1}	**10^0**	10^3	10^6	10^7
μm	10^{-9}	10^{-6}	10^{-4}	10^{-3}	**10^0**	10^3	10^4
nm	10^{-12}	10^{-9}	10^{-7}	10^{-6}	10^{-3}	**10^0**	10^1
Å[b]	10^{-13}	10^{-10}	10^{-8}	10^{-7}	10^{-4}	10^{-1}	**10^0**

[a] See instructions above.
[b] Not an SI unit.

Table C6.   Conversion Factors for Mole-Related Units[a]

Source	Target					
	mol	*mmol*	*μmol*	*nmol*	*pmol*	*fmol*
mol	**10^0**	10^3	10^6	10^9	10^{12}	10^{15}
mmol	10^{-3}	**10^0**	10^3	10^6	10^9	10^{12}
μmol	10^{-6}	10^{-3}	**10^0**	10^3	10^6	10^9
nmol	10^{-9}	10^{-6}	10^{-3}	**10^0**	10^3	10^6
pmol	10^{-12}	10^{-9}	10^{-6}	10^{-3}	**10^0**	10^3
fmol	10^{-15}	10^{-12}	10^{-9}	10^{-6}	10^{-3}	**10^0**

[a] To use with molarity (M), equivalence (equiv) or normality (N), substitute mol with M, equiv or N.

Appendix D

Commercial Concentrated
Acids and Bases

Table D1. Concentrations of Commercial Concentrated Acids and Bases

Acid or Base	Mol. weight	% (w/w)	Sg[a]	Approx. molarity	Approx. normality
Acetic acid	60.05	99.7	1.05	17.43	17.4
Ammonium hydroxide	35.05	28	0.90	14.8	14.8
Formic acid	46.03	97	1.22	25.7	25.7
Hydrochloric acid	36.46	37	1.20	12.18	12.1
Lactic acid	90.08	85	1.21	11.42	11.4
Nitric acid	63.01	70	1.40	15.55	15.5
Perchloric acid	100.46	70	1.66	11.57	11.6
Phosphoric acid	98.0	85	1.69	14.66	44.1
Sulfuric acid	98.1	95	1.84	17.82	35.6

[a] Specific gravity

Table D2. Volumes of Commercial Concentrated Acids and Bases Needed to Prepare Dilute Solutions

Acid or Base[a]	Milliliters required to prepare 1 liter of the following solutions		
	0.1 N	0.5 N	1.0 N
Acetic acid	5.7	28.7	57.4
Ammonium hydroxide	6.8	33.8	67.6
Formic acid	3.9	19.5	38.9
Hydrochloric acid	8.2	41.1	82.1
Lactic acid	8.7	43.8	87.6
Nitric acid	6.4	32.2	64.3
Perchloric acid	8.6	43.3	86.4
Phosphoric acid	2.3	11.4	22.7
Sulfuric acid	2.8	14.0	28.1

[a] See Table D-1 for concentrations

Appendix E

Table E1. pK$_a$ Values of Common Weak Acids

Free acid[a]	Anhydrous MW	pK$_{a_1}$	pK$_{a_2}$	pK$_{a_3}$
Acetic acid	60.05	4.76	—	—
Ascorbic acid	176.14	4.10	1.80	—
Barbituric acid	128.08	5.00		
Benzoic acid	122.13	4.19		
Butyric acid	88.12	4.82		
Boric acid	61.80	9.23		
CAPS	221.32	10.40	—	—
Carbonic acid	62	6.10	10.25	—
Citric acid	192.12	3.09	4.75	5.40
EDTA	292.24	1.70	2.60	6.30
EPPS	252.33	8.0	—	—
Glycine	75.07	2.40	9.6	—
Glycylglycine	132.12	3.10	8.4	—
Formic acid	46.03	3.74	—	
HEPES	238.31	7.55	—	—
Histidine	155.16	1.82	6.0	9.17
Lactic acid	90.05	3.86	—	
Imidazole	68.08	7.00	—	—
MES	195.20	6.15	—	—
MOPS	209.26	7.2	—	—
Nicotinic acid	123.11	4.85		
MOPSO	225.27	6.9	—	—
Oxalic acid	90.04	1.18	4.22	
PIPES	302.37	3.0	6.8	—
Phosphoric acid	98.00	1.96	6.8	12.32
Propionic acid	74.09	4.85		
Phthalic acid	166.13	2.90	5.4	—
Succinic acid	118.09	4.19	5.55	—
TAPS	243.28	8.4	—	—
TES	229.25	7.5	—	—
TRICINE	180.18	8.15	—	—
TRIS	121.14	8.10	—	—
Veronal acid	184.20	7.43	—	

[a] Trivial names used are defined as follows: BICINE, N,N-bis(2-hydroxyethyl) glycine; CAPS, 3-cyclohexylamino-1-propanesolfonic acid; EDTA, ethylenediaminetetraacetic acid; EPPS, 4-(2-hydroxyethyl)-1-piperazinepropanesulfonic acid; HEPES, 4-(2-hydroxyethyl-1-poperazineetane sulfonic acid; MES, 4-morpholineethanesulfonic acid; MOPS, 4-morpholine propanesulfonic acid; MOPSO, β-hydroxy-4-morpholinepropanesulfonic acid; PIPES, 1,4-piperazine(ethanesulfonic acid); TAPS, 3-methylamino-1-propanesulfonic acid; TRICINE, N-tris(hydroxymethyl)methylglycine; TRIS,(hydroxymethyl) aminomethane.

Table E2. pK_b Values of Common Weak Bases

Free base	Anhydrous MW	pK_{b_1}	pK_{b_2}	pK_{b_3}
Ammonia	17.03	4.74	—	—
Aniline	93.13	9.42	—	—
Diethylamine	73.14	3.02	—	—
Ethylamine	45.09	3.25	—	—
Hydrazine	32.04	5.77	—	—
Hydroxylamine	33.03	7.97	—	—
Methylamine	31.06	3.43	—	—
Triethylamine	101.19	3.30	—	—
Pyridine	79.10	8.77	—	—

Table E3. pH Values of Common Liquids

Liquid	pH Range
Human gastric juices	1.0–3.0
Lemon juice	2.2–2.4
Vinegar	2.4–3.4
Carbonated drinks	2.0–4.0
Orange juice	3.0–4.0
Black coffee	3.7–4.1
Tomato juice	4.0–4.4
Milk (cow's)	6.3–6.6
Human blood	7.3–7.5
Seawater	7.8–8.3
Ammonia (household)	10.5–11.5

Appendix F

Table F1. Useful Data for Nucleic Acids[a]

Nucleotide	Anhydrous MW[b]	λ_{max} (nM)	$\varepsilon_{\lambda max}$ (pH 7)	ε_{260nm} (pH 7)
	(Free acid form)			
Adenosine Triphosphate	507	259	15.4	15.0
Guanosine Triphosphate	523	253	13.7	11.8
Cytidine Triphosphate	483	271	9.1	7.4
Thymidine Triphosphate	498	267	9.6	8.4
Uridine Triphosphate	484	260	10.0	9.0
Inosine Triphosphate	508	249	12.2	7.4

Approximate MW of single-stranded DNA or RNA ≈ number of nucleotides X 325. Approximate MW of double-stranded DNA or RNA ≈ number of nucleotides X 650. A_{260} /A_{280} is 1.8 for pure DNA, and 2.0 for pure RNA. Lower values indicate contamination. 1 Absorbance unit is 50 μg/ml for double stranded DNA, 40 μg/ml for single-stranded DNA or RNA, 33 μg/ml for oligonucleotides with 40-100 bases, and 25 μg/ml for oligonucleotides less than 40 bases long.

[a] See also Table 5.1.

[b] To obtain the MW of the diphosphates, monophosphates and unphosphorylated, subtract 80 from each missing phosphate. The spectral data given apply to the phosphorylated forms also, except in occassional cases where λ(max) differ slightly.

Appendix G

Table G1. Symbols, Molecular Weight, and pK$_a$ Values of α-Amino Acids

Amino acid	3-Letter symbol	1-Letter symbol	MW wt.	pK$_a$ of:		
				α-COOH	α-NH$_3^+$	-R group
Alanine	Ala	A	89	2.35	9.69	—
Arginine	Arg	R	174	2.17	9.04	12.48
Asparagine	Asn	N	132	2.02	8.8	—
Aspartic acid	Asp	D	133	2.09	9.82	3.86
Cysteine	Cys	C	121	1.71	10.78	8.33
Glutamic acid	Glu	E	147	2.19	9.67	4.25
Glutamine	Gln	Q	146	2.17	9.13	—
Glycine	Gly	G	75	2.34	9.6	—
Histidine	His	H	155	1.82	9.17	6.0
Isoleucine	Ile	I	131	2.36	9.68	—
Leucine	Leu	L	131	2.36	9.60	—
Lysine	Lys	K	146	2.18	8.95	10.53
Methionine	Met	M	149	2.28	9.21	—
Phenylalanine	Phe	F	165	1.83	9.13	—
Proline	Pro	P	115	1.99	10.60	—
Serine	Ser	S	105	2.21	9.15	—
Threonine	Thr	T	119	2.63	10.43	—
Tryptophan	Trp	W	204	2.38	9.36	—
Tyrosine	Tyr	Y	181	2.20	9.11	10.07

Appendix H

Recipe for Preparing Common
Laboratory Buffers

1.   Phosphate Buffer

To prepare phosphate buffer at any concentration (X molar), first prepare X molar solutions of: (1) HPO_4^{2-} using K_2HPO_4 or Na_2HPO_4, and (2) $H_2PO_4^-$ using KH_2PO_4 or NaH_2PO_4. Prepare 100 ml, 300 ml and 1 L of the buffer following the table below. Check the pH with a pH electrode: If a slight adjustment is needed, add concentrated acid or base dropwise. **Note: $[HPO_4^{2-}]$ and $[H_2PO_4^-]$ must be identical and equal to that of the buffer to be prepared.** See problem 88d.

2.   Tris-HCl Buffer

To prepare Tris-HCl buffer at any concentration (X molar), first prepare X molar solutions of: (1) Tris base and (2) HCl. Prepare 100 ml, 300 ml, and 1 L of the buffer following the table below. Check the pH with a pH electrode. If a slight adjustment is needed, add concentrated acid or base dropwise. **Note: [Tris base] in solution 1, and [HCl] in solution 2 must be identical and equal to that of the buffer to be prepared.** See problem 88.

3.   Acetate Buffer

To prepare acetate buffer at any concentration (x molar), first prepare x molar solutions of: (1) acetate using sodium or potassium acetate, and (2) acetic acid. Prepare 100 ml, 300 ml and 1 L of the buffer following the table below. Check the pH with a pH electrode: If a slight adjustment is needed, add concentrated acid or base dropwise. **Note: [acetate] in solution 1 and [acetic acid] in solution 2 must be identical and equal to that of the buffer to be prepared.** See problem 88.

Table H1.   Phosphate Buffer

| | To prepare | | | | | |
| | 100 ml buffer | | 300 ml buffer | | 1000 ml buffer | |
pH	Add this x ml of x molar HPO_4^{2-}	To this ml of x molar $H_2PO_4^-$	Add this x ml of x molar HPO_4^{2-}	To this ml of x molar $H_2PO_4^-$	Add this ml of x molar HPO_4^{2-}	To this ml of x molar $H_2PO_4^-$
6.3	24.0	76.0	72.0	228.0	240.0	760.0
6.5	33.4	66.6	100.2	199.8	333.9	666.1
6.7	44.3	55.7	132.8	167.2	442.7	557.2
6.8	50.0	50.0	150.0	150.0	500.0	500.0
6.9	55.7	44.3	167.2	132.8	557.3	442.7
7.0	61.3	38.7	183.0	116.1	613.1	386.9
7.2	71.5	28.5	214.6	85.4	715.3	284.7
7.4	79.9	20.1	239.8	60.2	799.2	200.8
7.5	83.4	16.6	250.1	49.9	833.7	166.3
7.8	90.9	9.1	272.7	27.3	909.1	90.9

Table H2. Tris-HCl Buffer

pH	100 ml buffer		300 ml buffer		1000 ml buffer	
	Add this ml of x molar Tris base	To this ml of x molar HCl	Add this ml of x molar Tris base	To this ml of x molar HCl	Add this ml of x molar Tris base	To this ml of x molar HCi
7.4	16.6	83.4	49.9	250.1	166.3	833.7
7.5	20.1	79.9	60.2	239.8	200.8	799.2
7.6	24.0	76.0	72.1	227.9	240.3	759.7
7.7	28.5	71.5	85.4	214.6	284.7	715.3
7.8	33.4	66.6	100.2	199.8	333.9	666.1
7.9	38.7	61.3	116.1	183.9	386.9	613.1
8.0	44.3	55.7	132.8	167.2	442.7	557.3
8.1	50.0	50.0	150.0	150.0	500.0	500.0
8.2	55.7	44.3	167.2	132.8	557.3	442.7
8.3	61.3	38.7	183.9	116.1	613.1	386.9
8.4	66.6	33.4	199.8	100.2	666.1	333.9
8.5	71.5	28.5	214.6	85.4	715.3	284.7
8.6	76.0	24.0	227.9	72.1	759.7	240.3
8.8	83.4	16.6	250.1	49.9	833.7	166.3

Table H3. Acetate Buffer

	To prepare					
	100 ml buffer		300 ml buffer		1000 ml buffer	
pH	Add this ml of x molar acetate	To this ml of x molar acetic acid	Add this ml of x molar acetate	To this ml of x molar acetic acid	Add this ml of x molar acetate	To this ml of x molar acetic acid
4.0	13.7	86.3	41.0	259.0	136.8	863.2
4.2	20.1	79.9	60.2	239.8	200.8	799.2
4.4	28.5	71.5	85.4	214.6	284.7	715.3
4.5	33.4	66.6	100.2	199.8	333.9	666.1
4.6	38.7	61.3	116.1	183.9	386.9	613.1
4.7	44.3	55.7	132.8	167.2	442.7	557.3
4.8	50.0	50.0	150.0	150.0	500.0	500.0
4.9	55.7	44.3	167.2	132.8	557.3	442.7
5.0	61.3	38.7	183.9	116.1	613.1	386.9
5.1	66.6	33.4	199.8	100.2	666.1	333.9
5.2	71.5	28.5	214.6	85.4	715.3	284.7
5.3	76.0	24.0	227.9	72.1	759.7	240.3
5.4	79.9	20.1	239.8	60.2	799.2	200.8
5.6	86.3	13.7	259.0	41.0	863.2	136.8

Table H4.   Common Laboratory Buffers

Buffer	Concentration of 1x Solution	Recipe and Instructions for Preparing 1 L Solution
Tris-EDTA	10 mM Tris-Cl, 1 mM EDTA	**1× solution:** 1.21 g Tris base, 2 ml of 0.5 M EDTA **10× solution:** 12.11 g Tris base, 20 ml of 0.5 M EDTA Dissolve the Tris in 800 ml deionized water. Adjust the pH to desired value with 10 M Hcl. Q.s. To 1 L.
Tris-Acetate-EDTA (pH 7.8)	40 mM Tris-Acetate, 1 mM EDTA	**1× solution:** 4.84 g Tris base, 1.53 ml conc. Glacial acetic acid, 2 ml of 0.5 M EDTA. **10× solution:** 48.44 g Tris base, 15.29 ml conc. Glacial acetic acid, 2 ml of 0.5 M EDTA. Dissolve the Tris in 800 ml deionized water. While stirring, add the EDTA and acetic acid (slowly). Check the pH, and if minor adjustment is needed, use acetic acid or tris base. Q.s. To 1L.
Tris-Borate-EDTA (pH 8.3)	90 mM Tris-Borate, 1 mM EDTA	**0.5×:** 5.45 g Tris base, 3.41 g boric acid, 2 ml of 0.5 M EDTA. **1×:** 10.9 g Tris base, 6.82 g boric acid, 2 ml of 0.5 M EDTA. **10×:** 109 g Tris base, 68.2 g boric acid, 20 ml 0.5 M EDTA. Dissolve the Tris and boric acid in 800 ml deionized water. Add the EDTA. Check the pH, and if minor adjustment is needed, use boric acid or tris base. Q.s. To 1L.
Tris-Glycine (pH 8.3)	25 mM Tris, 203 mM glycine	**1×:** 3.03 g Tris base, 15.2 g glycine. **10×:** 30.3 g Tris base, 152.0 g glycine. Dissolve the Tris and glycince in 800 ml deionized water. Check the pH, and if minor adjustment is needed, use Tris base or glycine. Q.s. To 1 L.
Tris-Glycine-SDS (pH 8.3)	25 mM Tris base, 203 mM glycine, 0.1% sodium dodecyl sulfate (SDS)	**1×:** 3.03 g Tris base, 15.2 g glycine, 1 g sodium dodecyl sulfate. **10×:** 30.3 g Tris base, 152.0 g glycine, 10 g sodium dodecyl sulfate. Dissolve the Tris and glycince in 800 ml deionized water. Check the pH, and if minor adjustment is needed, use Tris base or glycine. Add and dissolve the SDS. Q.s. To 1 L.
Sample Buffer for PAGE and SDS-PAGE (pH 6.8)	62.3 mM Tris base, 10% glycerol, 2% SDS, 140 mM β-mercaptoethanol, 0.01% bromphenol blue	**Amounts given per 100 ml solution.** **2×:** 1.51 g Tris base, 20 ml glycerol, 4 g SDS, 2 ml β-mercaptoethanol, 0.02 g bromphenol blue. **5×:** 3.78 g Tris base, 50 ml glycerol, 10 g SDS, 5 ml β-mercaptoethanol, 0.05 g bromphenol blue. Dissolve the Tris in 30 ml deionized water, add and dissolve the glycerol. Adjust the pH to 6.8 with conc. HCl. Add and dissolve the SDS, β-mercaptoethanol, bromphenol blue, and q.s. to 1 L. Store at 25°C.

Appendix I

Table I1. Half-Life of Selected Isotopes[a]

Isotope	Half-Life[b]	Type of particle emitted	Major decay energies (MeV)
Carbon-14 (^{14}C)	5760 years	β^-	0.155
Hydrogen-3 (^3H)	12.26 years	β^-	0.018
Iodine-125 (^{125}I)	60 days	EC	
Iodine-131 (^{131}I)	8.04 days	β^-	0.61, 0.33
		γ	0.36, 0.64
Phosphorus-32 (^{32}P)	14.2 days	β^-	1.71
Phosphorus-33 (^{33}P)	25 days	β^-	0.25
Potassium-40 (^{40}K)	1.3×109 years	β^-	1.32
		γ	1.46
		EC	—
Rubidium-86 (^{86}Rb)	18.7 days	β^-	1.82
		γ	1.1
Sodium-22 (^{22}Na)	2.6 years	β^+	0.55, 0.58, 1.8
		γ	0.51, 1.27
Sulfur-35 (^{35}S)	87.2 days	β^-	0.167

[a] See p. 125 for radioactivity units.

[b] See Source; ref 12.

Appendix J

Table J1. Fractions of ^{32}P and ^{131}I Remaining over Time

Time elapsed (days)	Radioactivity remaining (fraction)	
	^{32}P	^{131}I
0.5	0.976	0.958
1.0	0.953	0.918
1.5	0.930	0.880
2.0	0.908	0.843
2.5	0.886	0.807
3.0	0.865	0.774
3.5	0.844	0.741
4.0	0.824	0.710
4.5	0.804	0.681
5.0	0.785	0.652
5.5	0.766	0.625
6.0	0.748	0.599
6.5	0.730	0.573
7.0	0.712	0.549
7.5	0.695	0.526
8.0	0.679	0.504
9.0	0.647	0.463
10.0	0.616	0.425
11.0	0.587	0.390
12.0	0.559	0.358
13.0	0.533	0.329
14.0	0.507	0.302
15.0	0.483	0.277
16.0	0.461	0.254
16.0	0.461	0.254

Table J2.   Fractions of ^{125}I and ^{35}S Remaining over Time

Time elapsed (days)	Radioactivity remaining (fraction)	
	^{125}I	^{35}S
2	0.977	0.984
4	0.955	0.969
6	0.933	0.953
8	0.912	0.938
10	0.891	0.923
12	0.871	0.909
14	0.851	0.895
16	0.831	0.881
18	0.812	0.869
20	0.794	0.853
22	0.776	0.839
24	0.758	0.826
26	0.741	0.813
28	0.724	0.800
30	0.707	0.788
32	0.691	0.775
34	0.675	0.763
36	0.660	0.751
38	0.645	0.739
40	0.630	0.727
42	0.616	0.716
44	0.602	0.705
46	0.588	0.694
48	0.574	0.683
50	0.561	0.672
52	0.549	0.661
54	0.536	0.651
56	0.524	0.641
60	0.500	0.620
64	0.478	0.601
68	0.456	0.582
72	0.435	0.564
76	0.416	0.546
80	0.397	0.529
84	0.379	0.513
88	0.362	0.497

Appendix K

Table K1. Atomic Weights of Selected Elements		Table K1. Atomic Weights of Selected Elements	
Element (symbol)	**Atomic weight**	**Element (symbol)**	**Atomic weight**
Aluminum (Al)	26.98	Mercury (Hg)	200.59
Antimony (Sb)	121.75	Molybdenum (Mo)	95.94
Argon (Ar)	39.95	Neon (Ne)	20.18
Arsenic (As)	74.92	Nickel (Ni)	158.71
Barium (Ba)	137.34	Niobium (Nb)	92.91
Beryllium (Be)	9.01	Nitrogen (N)	14.01
Bismuth (Bi)	208.98	Osmium (Os)	190.2
Boron (B)	79.91	Oxygen (O)	16.00
Bromine(Br)	79.90	Palladium (Pd)	106.4
Cadmium (Cd)	112.40	Phosphorus (P)	30.97
Calcium (Ca)	40.08	Platinum (Pt)	195.09
Carbon (C)	12.01	Potassium (K)	39.10
Cesium (Cs)	132.90	Radon (Rn)	222
Chlorine (Cl)	35.45	Rhodium (Rh)	102.90
Chromium (Cr)	52.00	Rubidium (Rb)	85.47
Cobalt (Co)	58.93	Selenium (Se)	78.96
Copper (Cu)	63.54	Silicon (Si)	28.09
Fluorine (F)	19.00	Silver (Ag)	107.78
Gold (Au)	197.00	Sodium (Na)	23.00
Helium (He)	4.00	Sulfur (S)	32.06
Hydrogen (H)	1.01	Thallium (Ti)	204.37
Indium (In)	114.82	Tin (Sn)	118.69
Iodine (I)	126.90	Titanium (Ti)	47.90
Iridium (Ir)	192.2	Tungsten (W)	183.85
Iron (Fe)	55.85	Uranium (U)	238.03
Lanthanum (La)	138.91	Vanadium (V)	50.94
Lead (Pb)	207.19	Xenon (Xe)	131.30
Lithium (Li)	6.94	Ytterbium (Yb)	173.04
Magnesium (Mg)	24.31	Zinc (Zn)	65.37
Manganese (Mn)	54.94		

Appendix L

Useful Constants and Values for Colligative Properties of Solutions

Table L1. Vapor Pressure of Water as a Function of Temperature

Temp. (EC)	Pressure (torr)	Temp. (EC)	Pressure (torr)	Temp. (EC)	Pressure (torr)
0	4.6	18	15.5	40	55.3
1	4.9	19	16.5	45	71.9
2	5.3	20	17.5	50	92.5
3	5.7	21	18.7	55	118.0
4	6.1	22	19.8	60	149.4
5	6.5	23	21.1	65	187.5
6	7.0	24	22.4	70	233.7
7	7.5	25	23.8	75	289.1
8	8.0	26	25.2	80	355.1
9	8.6	27	26.7	85	433.6
10	9.2	28	28.3	90	525.8
11	9.8	29	30.0	95	634.1
12	10.5	30	31.8	96	657.6
13	11.2	31	33.7	97	682.1
14	12.0	32	35.7	98	707.3
15	13.1	33	37.7	99	733.2
16	13.6	34	39.9	100	760.0
17	14.5	35	42.2	101	787.6

Table L2. Molal Boiling Point Elevation Constants (K_b) and Freezing Point Depression Constants (K_f) for Some Solvents

Solvent	Boiling point (°C)	K_b (°C·Kg·mol^{-1})	Freezing point (°C)	K_f (°C·Kg·mol^{-1})
Water	100.0	0.52	0	1.86
Carbon tetrachloride	76.5	5.03	-22.99	30
Chloroform	61.2	3.63	-63.5	4.70
Benzene	80.1	2.53	5.5	5.12
Carbon disulfide	46.2	2.34	-111.5	3.83
Ethyl ether	34.5	2.02	-116.2	1.79
Camphor	208.0	5.95	179.8	40
Acetic acid	118.2	2.93	17	3.90

Appendix M

Description of Spreadsheet Contents

In the following tables, only those cells that contain references, calculations, or other formulas are listed. Those cells that contain descriptions, titles, or other types of placeholders can be found in the examples within the text.

Table M1.   Table of Cell Descriptions for Molecular Calculator Spreadsheet

CELL	FORMULA	CELL	FORMULA
C26	SUM(C8:C24)	D32	$D$30*D268$D$31
D8	IF(C8="","",B8*C8)	E8	IF(C8="","",(D8/$D$31*100))
D9	IF(C9="","",B9*C9)	E9	IF(C9="","",(D9/$D$31*100))
D10	IF(C10="","",B10*C10)	E10	IF(C10="","",(D10/$D$31*100))
D11	IF(C11="","",B11*C11)	E11	IF(C11="","",(D11/$D$31*100))
D12	IF(C12="","",B12*C12)	E12	IF(C12="","",(D12/$D$31*100))
D13	IF(C13="","",B13*C13)	E13	IF(C13="","",(D13/$D$31*100))
D14	IF(C14="","",B14*C14)	E14	IF(C14="","",(D14/$D$31*100))
D15	IF(C15="","",B15*C15)	E15	IF(C15="","",(D15/$D$31*100))
D16	IF(C16="","",B16*C16)	E16	IF(C16="","",(D16/$D$31*100))
D17	IF(C17="","",B17*C17)	E17	IF(C17="","",(D17/$D$31*100))
D18	IF(C18="","",B18*C18)	E18	IF(C18="","",(D18/$D$31*100))
D19	IF(C19="","",B19*C19)	E19	IF(C19="","",(D19/$D$31*100))
D20	IF(C20="","",B20*C20)	E20	IF(C20="","",(D20/$D$31*100))
D21	IF(C21="","",B21*C21)	E21	IF(C21="","",(D21/$D$31*100))
D22	IF(C22="","",B22*C22)	E22	IF(C22="","",(D22/$D$31*100))
D23	IF(C23="","",B23*C23)	E23	IF(C23="","",(D23/$D$31*100))
D24	IF(C24="","",B24*C24)	E24	IF(C24="","",(D24/$D$31*100))
D26	SUM(D8:D24)	E26	SUM(E8:E24)

Table M2.   Table of Cell Descriptions for pH Titration Graph

CELL	FORMULA
E10	=$G$6-$G$7
E11	=$G$5-$G$6
E13	=IF(AND($G$11>$F$13,$F$13>$G$10),"Equivalence point pH: ","Wrong Volume Choosen")
E5	=VLOOKUP($G$3,$A$3:$B$16,2,FALSE)
E6	=OFFSET($B$2,($F$5-1))
E7	=OFFSET($B$2,($F$5-2))
E8	=OFFSET($B$2,($F$5-3))
F13	=IF(AND($E$11<0,$E$10>0),-(((($G$10-$G$11)*$E$10)/($E$10-$E$11))-$G$10),"")
F5	=MATCH($E$4,$B$3:$B$16)
G10	=$E$6
G11	=$E$5
G3	User defined volume
G5	=$E$5-$E$6
G6	=$E$6-$E$7
G7	=$E$7-$E$8

Table M3. Table of Cell Descriptions for Dilution Preparation Worksheet

CELL	FORMULA
C15	=IF($C$24>1,"Too many X's",IF($D$28<>2,"",IF($B$7="X",($D8*$E8/$C8),IF($C$7="X",($D8*$E8/$B8),
(con't)	IF($D$7="X",($B8*$C8/$E8),IF($E$7="X",($B8*$C8/$D8),""))))))
C20	=IF($B$7="X", 1, 0)
C21	=IF($C$7="X", 1, 0)
C22	=IF($D$7="X", 1, 0)
C23	=IF($E$7="X", 1, 0)
C24	=SUM(C20:C23)
C26	=IF($B$6=$D$6,0,1)
C27	=IF($C$6=$E$6,0,1)
C28	=SUM(C26:C27)
D15	=IF($E$28<>2,"",(IF($B$7="X",$B$6,IF($C$7="X",$C$6,IF($D$7="X",$D$6,IF($E$7="X",$E$6,""))))))
D26	=IF($E$28<>2,0,1)
D27	=IF($C$24>1,0,1)
D28	=SUM(D26:D27)
E20	=IF($B$6="", 1, 0)
E21	=IF($C$6="", 1, 0)
E22	=IF($D$6="", 1, 0)
E23	=IF($E$6="", 1, 0)
E24	=SUM(E20:E23)
E26	=IF($E$24>0,0,1)
E27	=IF($C$28>0,0,1)
E28	=SUM(E26:E27)

Table M4. Table of Cell Descriptions for Routine Calculation Template

CELL	FORMULA
D18	=($C$13*0.001*$C$12*$D18)/(($B18/$C18)/$C$14)
D19	=($C$13*0.001*$C$12*$D19)/(($B19/$C19)/$C$14)
D20	=($C$13*0.001*$C$12*$D20)/(($B20/$C20)/$C$14)
D21	=($C$13*0.001*$C$12*$D21)/(($B21/$C21)/$C$14)
D22	=($C$13*0.001*$C$12*$D22)/(($B22/$C22)/$C$14)
D23	=($C$13*0.001*$C$12*$D23)/(($B23/$C23)/$C$14)
D24	=($C$13*0.001*$C$12*$D24)/(($B24/$C24)/$C$14)
D25	=($C$13*0.001*$C$12*$D25)/(($B25/$C25)/$C$14)

Table M5. Table of Cell Descriptions for Calibration Regression Analysis

CELL	FORMULA	CELL	FORMULA	CELL	FORMULA	CELL	FORMULA
B11	=SUM(B6:B10)	D8	=POWER(B8,2)	B28	=$B$6	G28	=$B$6
B13	=$D$11-(POWER(B11,2)/B12)	D9	=POWER(B9,2)	B29	=$B$7	G29	=$B$7
B14	=$E$11-(POWER(C11,2)/B12)	E10	=POWER(C10,2)	B30	=$B$8	G30	=$B$9
B15	=$F$11-((B11*C11)/B12)	E11	=SUM(E6:E10)	B31	=$B$9	G31	=$B$10
B16	=B15/B13	E46	=POWER(C46,2)	B32	=$B$10	G34	=$B$6
B17	=(C11/B12)-B16*(B11/B12)	E47	=POWER(C47,2)	B34	=$B$7	G35	=$B$7
B21	=RSQ(B28:B32,C28:C32)	E48	=POWER(C48,2)	B35	=$B$8	G36	=$B$8
B22	=RSQ(B34:B37,C34:C37)	E49	=POWER(C49,2)	B36	=$B$9	G37	=$B$10
B23	=RSQ(B39:B42,C39:C42)	E50	=SUM(E46:E49)	B37	=$B$10	G39	=$B$6
B24	=RSQ(G28:G31,H28:H31)	E6	=POWER(C6,2)	B39	=$B$6	G40	=$B$7
B25	=RSQ(G34:G37,H34:H37)	E7	=POWER(C7,2)	B40	=$B$8	G41	=$B$8
B26	=RSQ(G39:G42,H39:H42)	E8	=POWER(C8,2)	B41	=$B$9	G42	=$B$9
B50	=SUM(B46:B49)	E9	=POWER(C9,2)	B42	=$B$10	H28	=$C$6
B52	=$D$50-(POWER(B50,2)/B51)	F10	=PRODUCT(B10,C10)	C28	=$C$6	H29	=$C$7
B53	=$E$50-(POWER(C50,2)/B51)	F11	=SUM(F6:F10)	C29	=$C$7	H30	=$C$9
B54	=$F$50-((B50*C50)/B51)	F46	=PRODUCT(B46,C46)	C30	=$C$8	H31	=$C$10
B55	=B54/B52	F47	=PRODUCT(B47,C47)	C31	=$C$9	H34	=$C$6
B56	=(C50/B51)-B55*(B50/B51)	F48	=PRODUCT(B48,C48)	C32	=$C$10	H35	=$C$7
C11	=SUM(C6:C10)	F49	=PRODUCT(B49,C49)	C34	=$C$7	H36	=$C$8
C50	=SUM(C46:C49)	F50	=SUM($F46:$F49)	C35	=$C$8	H37	=$C$10
D10	=POWER(B10,2)	F6	=PRODUCT(B6,C6)	C36	=$C$9	H39	=$C$6
D11	=SUM(D6:D10)	F7	=PRODUCT(B7,C7)	C37	=$C$10	H40	=$C$7
D46	=POWER(B46,2)	F8	=PRODUCT(B8,C8)	C39	=$C$6	H41	=$C$8
D47	=POWER(B47,2)	F9	=PRODUCT(B9,C9)	C40	=$C$8	H42	=$C$9
D48	=POWER(B48,2)	H20	=IF($B$23>=0.995, "YES", "NO")	C41	=$C$9	H59	=$H$22
D49	=POWER(B49,2)	H21	=IF($H$20="YES", $B$21, MAX(B22:B26))	C42	=$C$10	H68	=$H$21
D50	=SUM(D46:D49)	H22	=VLOOKUP($H$21,B21:C26,2,FALSE)				
D6	=POWER(B6,2)	H69	=IF($H$20="YES",$B$16, $B$55)				
D7	=POWER(B7,2)	H70	=IF($H$20="YES",$B$17, $B$56)				

Table M6.   MACRO Program Details for Section 8.2F, Figures 8.8 and 8.9

```
Sub Macro1 ( )
'
' Graphing Macro
' Written 10/8/00 by Todd A. Jones
'
' Range 'Data_array' is defined as cell H22.
' Range 'all_A' is defined as cells B34:C37.
' Range 'all_B' is defined as cells B34:C42.
' Range 'all_C' is defined as cells G28:H31.
' Range 'all_D' is defined as cells G34:G37.
' Range 'all_E' is defined as cells G39:G42.
' Range 'all_5_points' is defined as cells A60:B63.
' Range `blank1' is defined as cells A60:B63.
' Range `blank2' is defined as cells A64:B68.
'
' Matches data array selection in the data pair selection box. The correct
' data array is selected and placed in the secondary least-squares standard
' curve calculations. The same data array is also placed in the graphing box.

Range ("F62:G66"). ClearContents                          'Clears previous graph points.

If Range (Data_array") = 'all_B' Then                     'If cell H22 = all A, then
    Application.Goto Reference: ="all_A"                   'copy cells B34:C37 and
    Selection.Copy                                        'paste in cells B46 and F62.
    Range ("B46"). Select
    ActiveSheet.PasteSpecial
    Range ("F62"). Select
    ActiveSheet.PasteSpecial
ElseIf Range ("Data_array") = "all_B" Then
    Application.Goto Reference: ="all_B"
    Selection.Copy
    Range ("B46"). Select
    ActiveSheet.PasteSpecial
    Range ("F62"). Select
    ActiveSheet.PasteSpecial
ElseIf Range ("Data_array") = "all_C" Then
    Application.Goto Reference: ="all_C"
    Selection.Copy
    Range ("B46"). Select
    ActiveSheet.PasteSpecial
    Range ("F62"). Select
    ActiveSheet.PasteSpecial
ElseIf Range ("Data_array") = "all_D" Then
    Application.Goto Reference: ="all_D"
    Selection.Copy
    Range ("B46"). Select
    ActiveSheet.PasteSpecial
    Range ("F62"). Select
    ActiveSheet.PasteSpecial
ElseIf Range ("Data_array") = "all_E" Then
    Application.Goto Reference: ="all_E"
    Selection.Copy
    Range ("B62"). Select
    ActiveSheet.PasteSpecial
    Range ("F62"). Select
    ActiveSheet.PasteSpecial
ElseIf Range ("Data_array") = "All_5_points" Then        'Copy blank values into
    Application.Goto Reference: ="blank1"                 'secondary regression
    Selection.Copy                                        '(not used)
    Range ("B46"). Select
    ActiveSheet.PasteSpecial
    Application.Goto Reference:="All_5_points"            'Copy all five data pairs
    Selection.Copy                                        'into graphing cells.
    Range ("F62"). Select
    ActiveSheet.PasteSpecial
Else                                                      'ERROR: if H22 doesn't
    Application.Goto Reference:="blank"                   'contain any above options
    Selection.Copy                                        'paste three blanks values
    Range ("B46").Select                                  'in B46 and F62.
    ActiveSheet.PasteSpecial                              'NO DATA PAIRS in the graph
    Application.Goto Reference:"blank2"                   'indicates an error!!!
    Selection.Copy
    Range ("F62").Select
    ActiveSheet.PasteSpecial
    End If
Range ("A1").Select
End Sub
```

Index to Practical Examples

Page numbers are in parentheses.

Chapter 1

Chapter 2

Problems 28–32. (34–35) **Calculating percent composition of a compound** using molecular **mass data**.

Problems 33–37. (36–38) **Calculating empirical formula** using percent composition and weight data.

Problems 38–42. (39–41) **Calculating molecular formula** using **empirical formula, molar mass** and **% composition**.

Problem 43. (42) **Determining the limiting reactant** of a chemical reaction, using **reaction stoichiometry** data.
Calculating product yield.

Problem 44. (43) **Calculating product yield** using **reaction stoichiometry** data.

Problems 45–47. (44–45) Calculating product yield using **reaction stoichiometry** data.
Determining the limiting reactant of a chemical reaction, using **reaction stoichiometry** data.

Problems 48–50. (46–47) **Calculating % yield** of a chemical reaction using **reaction stoichiometry** data.

Problem 51. (48) **Calculating the theoretical, actual and % yields** of a chemical reaction, using **reaction stoichiometry** data.

Problems 52–54. (48–50) **Calculating chemical reaction order, rate, rate law and rate constant**.

Chapter 3

Problem 55. (59) **Calculating the moles of a gas,** using the **ideal gas law**.

Problems 56–57. (60) **Calculating changes in gas volume and pressure,** using **Boyle's law**.

Problems 58–59. (61) **Calculating changes in gas volume and temperature,** using **Charles' law**.

Problems 60–61. (62) **Calculating changes in gas pressure and temperature,** using **Gay Lussac's law**.

Problems 62–66. (63–65) **Calculating changes in gas pressure, volume and temperature,** using **the combined gas law**.

Problem 67. (65) **Calculating the moles and volume of a gas** produced in a reaction, using **Avogadro's law**.

Problems 68–70. (66–67) **Calculating gas pressure, volume and mole,** using the ideal **gas law**.

Problems 71–73. (67–68) **Calculating the grams, molecular weight and density of a gas,** using the **ideal gas law**.

Problems 74–77. (68–69) **Calculating the total pressure, partial pressure and mole fraction in a mixture of gases,** using the **Dalton's law of partial pressures**.

Problem 78. (70) **Calculating the mole fraction and molar mass of a solute in solution,** using **Raoult's law**.

Problems 79–83. (71–73) **Calculating the boiling point, boiling point elevation, freezing point and freezing point elevation of solvents, and the molecular weight and grams of solutes,** using colligative properties.

Problems 84–86. (73–74) **Calculating osmotic pressure, molecular weight and grams of solution and solutes,** using the principles of osmotic pressure.

Problem 87. (74) **Calculating the vapour pressure** over solvents.

Chapter 4

Problem 88. (88) **Calculating the amounts of conjugate acid and base needed to prepare a buffer** having specified pH, volume and concentration values.
Preparing common laboratory buffers by using exact weights of the appropriate conjugate acid and base, which when dissolved together, gives a buffer of predetermined pH, volume and concentration values.

Problem 89. (93) **Calculating the volumes of conjugate acid and base solutions needed to prepare a buffer** having specified pH, volume and concentration values.
Preparing common laboratory buffers, using calculated volumes of the solutions of the conjugate acid and base, which when combined, gives a buffer of given pH, volume and concentration values.

Problem 90. (95) **Calculating the moles of H$^+$ released into an assay medium** by biochemical reactions.

Index